VCE Units 3 & 4
GENERAL MATHEMATICS

MARK O'CONNELL
CHRISTINE McRAE

2023–2027 STUDY DESIGN • 2023–2027 STUDY DESIGN •

A+

+ topic summaries
+ graded exam practice questions
  with worked solutions
+ study and exam preparation advice

STUDY
NOTES

A+ VCE General Mathematics Study Notes
4th Edition
Mark O'Connell
Christine McRae
ISBN 9780170465335

Publisher: Cathy Beswick-Davison
Associate publisher: Naomi Campanale
Project editor: Tanya Smith
Editor: Karen Chin
Series text design: Nikita Bansal
Series cover design: Nikita Bansal
Series designer: Cengage Creative Studio
Artwork: MPS Limited
Production controller: Karen Young
Typeset by: Nikki M Group Pty Ltd

Any URLs contained in this publication were checked for currency during the production process. Note, however, that the publisher cannot vouch for the ongoing currency of URLs.

Acknowledgements
Selected VCE examination questions and extracts from the VCE Study Designs are copyright Victorian Curriculum and Assessment Authority (VCAA), reproduced by permission. VCE ® is a registered trademark of the VCAA. The VCAA does not endorse this product and makes no warranties regarding the correctness or accuracy of this study resource. To the extent permitted by law, the VCAA excludes all liability for any loss or damage suffered or incurred as a result of accessing, using or relying on the content. Current VCE Study Designs, past VCE exams and related content can be accessed directly at www.vcaa.vic.edu.au.

Casio ClassPad: Images used with permission by Shriro Australia Pty Ltd

For product information and technology assistance,
in Australia call **1300 790 853**;
in New Zealand call **0800 449 725**

For permission to use material from this text or product, please email **aust.permissions@cengage.com**

ISBN 978 0 17 046533 5

Cengage Learning Australia
Level 5, 80 Dorcas Street
Southbank VIC 3006 Australia

Cengage Learning New Zealand
Unit 4B Rosedale Office Park
331 Rosedale Road, Albany, North Shore 0632, NZ

For learning solutions, visit **cengage.com.au**

Printed in China by 1010 Printing International Limited.
1 2 3 4 5 6 7 26 25 24 23

CONTENTS

CHAPTER

1

UNIT 3

DATA ANALYSIS

Area of Study 1: Data analysis, probability and statistics

CHAPTER

2

RECURSION AND FINANCIAL MODELLING

Area of Study 2: Discrete mathematics

CHAPTER

3

CHAPTER

4

CHAPTER

5

HOW TO USE THIS BOOK

The *A+ VCE General Mathematics* resources are designed to be used year-round to prepare you for your VCE General Mathematics exam. *A+ VCE General Mathematics Study Notes* includes topic summaries of all key knowledge in the VCE General Mathematics Study Design. The first four chapters of this book addresses the Areas of Study of the course. The following gives you a brief overview of each chapter and the features included in this resource.

Area of Study summaries

A+ General Mathematics Study Notes includes topic summaries of the key knowledge in the VCE Mathematics Study Design accredited for 2023–2027, which is assessed in the final exam.

Concept maps

The concept map at the beginning of each chapter provides a visual summary of each Area of Study.

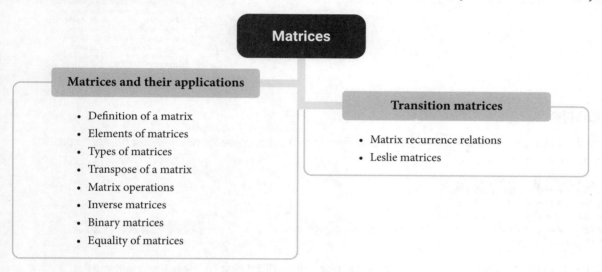

Exam practice

Exam practice questions appear at the end of each chapter to test you on what you have just reviewed in the chapter. These are written in the same style as the questions in the actual VCE General Mathematics exams. There are also official past VCAA exam questions in each chapter.

Multiple-choice questions

There are between 50 to 100 multiple-choice questions in each chapter.

Extended-answer questions

There are a wide range of extended-answer questions in each chapter. These questions are longer and often consist of parts that require you to apply your knowledge across single or multiple concepts. Mark allocations have been provided for each question.

9780170465335

Solutions

The last chapter of the book provides the solutions to all exam practice questions. Solutions to extended-answer questions have been written to reflect a high-scoring response and include explanations of what makes an effective answer.

Explanations

Where relevant, the solutions section includes explanations of multiple-choice answers. Extended-answer solutions outline what a high-scoring response looks like.

> **Question 73.1**
> a 200 km
> b *FDC, FEDC, FEBC, FEABC, FDEBC, FDEABC*
>   → 6

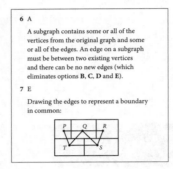

> 6 A
> A subgraph contains some or all of the vertices from the original graph and some or all of the edges. An edge on a subgraph must be between two existing vertices and there can be no new edges (which eliminates options **B, C, D** and **E**).
>
> 7 E
> Drawing the edges to represent a boundary in common:

> 60 C
> There are two paths (*AFH* and *BCFH*) of length 18 hours (the critical paths) and two paths (*AEG* and *BCEG*) of length 17 hours. Reducing activities *A* and *B* by 1 hour each reduces the lengths of these two paths to 17 and 16 hours respectively. Reducing either *F* or *H* by 1 hour will then further reduce the critical paths, *AFH* and *BCFH*, to 16 hours. Thus, a minimum of these activities must be reduced by 1 hour each to reduce the project completion time to 16 hours. It is not sufficient to just reduce the length of paths *AFH* and *BCFH* to 16 hours, for example, by reducing the durations of *F* and *H* by 1 hour each because it would still take 17 hours to complete the activities on the paths *AEG* and *BCEG*.

Icons

The icons below are found in the summaries and exam practice sections of each chapter to provide additional tips and support.

> **Note**
> In the long term, an investment, at the same rate, will earn less with simple interest than with compound interest.

Hint and note boxes appear throughout the summaries to provide additional tips and support.

©VCAA   2018N 2CQ9

This icon appears with official past VCAA exam questions.

These icons indicate the question's level of difficulty: easy, medium or hard.

A+ DIGITAL

Just scan the QR code or type the URL into your browser to access:

- A+ Flashcards: revise key terms and concepts online
- Revision summaries of all concepts from each extended-answer question.

Note: You will need to create a free NelsonNet account.

https://get.ga/aplus-vcegeneral-maths

PREPARING FOR EXAMS

Exam preparation is a year-long process. It is important to keep on top of the theory and consolidate often, rather than leaving study to the last minute. You should aim to have the theory learnt and your notes completed so that by the time you reach the pre-exam study period, the revision you do is structured, efficient and meaningful.

Study tips

To stay motivated, try to make the studying experience as comfortable as possible. Have a dedicated study space that is well lit and quiet. Create and stick to a study timetable, take regular breaks and reward yourself with social outings or treats.

Revision techniques

Here is a useful technique to help with your revision: **'STIC'** – a framework for structuring your learning so that you study less, but learn more!

Spaced repetition	This technique helps to move information from your short-term memory into your long-term memory by spacing out the time between your revision and recall flash card sessions. As the time between retrieving information is slowly extended, the brain processes and stores the information for longer periods.
Testing	Testing is necessary for learning and is a proven method for exam success. If you test yourself continually before you learn all the content, your brain becomes primed to retain important information when you learn it. As part of this process, engage with the marking criteria provided to help decide on the areas where improvement is needed.
Interleaving	This is a revision technique that sounds counterintuitive but is very effective for retaining information. Most students tend to revise a single topic in a session, and then move onto another topic in the next session. With interleaving, you choose three topics (1, 2, 3) and spend 20 to 30 minutes on each topic. You may choose to study 1-2-3 or 2-1-3 or 3-1-2, 'interleaving' the topics and repeating the study pattern over a long period of time. This strategy is most helpful if the topics are from the same subject and are closely related.
Chunking	An important strategy is breaking down large topics into smaller, more manageable 'chunks' or categories. Essentially, you can think of this as a branching diagram or mind map where the key theory or idea has many branches coming off it that get smaller and smaller. By breaking down the topics into these chunks, you will be able to revise the topic systematically.

These strategies take cognitive effort, but that is what makes them more effective than re-reading notes or trying to cram information into your short-term memory the night before the exam!

Time management

It is important to manage your time carefully throughout the year. Make sure you are getting enough sleep and the right nutrition, and that you are exercising and socialising to maintain a healthy balance.

To help you stay on target, plan your study timetable. Here is one way to do this:

1 Assess your current study time and social time. How much will you dedicate to each?

2 List all your commitments and deadlines, including sport, work, assignments, etc.

3 Prioritise the list and reassess your time, to ensure you can meet all your commitments.

4 Decide on a format, whether it be weekly or monthly, and schedule in a study routine.

5 Keep your timetable where you can see it.

6 Be consistent.

Studies suggest that one-hour blocks with a 10-minute break is most effective for studying, and remember you can interleave three topics during this time. You will also have free periods during the school day that you can use for study, note-taking, assignments, meeting with your teachers and group study sessions. Studying does not have to take hours if it is done effectively. Use your timetable to schedule short study sessions often.

Important information from the Study Design

Your school-based assessment will have addressed the skills required for working mathematically, which contributes to 40% of your study score.
The examinations will contribute 60 per cent to your study score. Each examination will contribute 30 per cent to your study score.

The exams

Examination 1

This examination comprises multiple-choice questions covering all areas of study. The examination is designed to assess students' knowledge of mathematical concepts, models and techniques and their ability to reason, interpret and apply this knowledge in a range of contexts.

Examination 2

This examination comprises written response questions covering all areas of study. The examination is designed to assess students' ability to select and apply mathematical facts, concepts, models and techniques to solve extended application problems in a range of contexts.

Conditions

The examinations will be completed under the following conditions:

- Duration: 1.5 hours

- Student access to an approved technology with numerical, graphical, symbolic, financial and statistical functionality will be assumed.

- One bound reference text (which may be annotated) or lecture pad, may be brought into the examination.

The following strategies will help you prepare for the exams.

Practise using past papers

To help prepare, download past papers from the VCAA website and attempt as many as you can in the lead-up to the exam. These will show you the types of questions to expect and provide practice in writing answers. It is a good idea to make the trial exams as much like the real exam as possible (conditions, time constraints, materials, etc.).

Use trial papers, school-assessed coursework and comments from your teacher to pinpoint weaknesses, and work on improving these areas. Do not just tick or cross your answers; look at the suggested answers and try to work out why your answer was different. What misunderstandings do your answers show? Are there gaps in your knowledge? Read the examiners' reports to find out the common mistakes students make.

Make sure you understand the material, rather than trying to rote learn information. Most questions are aimed at your understanding of concepts and your ability to apply your knowledge to new situations.

The day of the exam

The night before your exam, try to get a good rest and avoid cramming, as this will only increase stress levels. On the day of the exam, arrive at the venue early and bring everything you will need with you. If you must rush to the exam, your stress levels will increase, thereby lowering your ability to do well. Furthermore, if you are late you will have less time to complete the exam, which means you may not be able to answer all the questions or may rush to finish and make careless mistakes. If you are more than 30 minutes late, you may not be allowed to enter the exam. Do not worry too much about exam jitters. A certain amount of stress can help you to concentrate and achieve an optimum level of performance. If, however, you are feeling very nervous, breathe deeply and slowly. Breathe in for a count of 6 seconds, and out for 6 seconds until you begin to feel calm.

Reading time

Use your time wisely! Do not use the reading time to try and figure out the answers to any of the questions until you have read the whole paper. Plan your approach so that when you begin writing you know which section, and ideally which question, you are going to start with.

Strategies for answering Exam 1

Exam 1 consists of a question book and an answer sheet. The answers for multiple-choice questions must be recorded on the answer sheet provided. A correct answer scores 1, and an incorrect answer scores 0. There is no deduction for an incorrect answer, so attempt every question.

Read the question carefully and underline any important information to help you break the question down and avoid misreading it. Read all the possible solutions and eliminate any clearly wrong answers. You can annotate or write on any diagrams and make notes in the margins. Fill in the multiple-choice answer sheet carefully and clearly. Check your answer and move on. Do not leave any answers blank.

Strategies for answering Exam 2

Exam 2 consists of a question book with space to write your answers. The space provided is an indication of the detail required in the answer. Most questions will be broken down into several parts, and each part will be testing new information; so, read the entire question carefully to ensure you do not repeat yourself.

Make sure your handwriting is clear and legible. Use correct mathematical terminology. Attempt all questions: marks are not deducted for incorrect answers, and you might get some marks if you make an educated guess. You will definitely not get any marks if you leave a question blank!

The examiners' reports always highlight the importance of planning responses before writing. You have 1.5 hours to complete 60 marks. This means you have an average of 1.5 minutes per mark.

ABOUT THE AUTHORS

Christine McRae and Mark O'Connell

Christine McRae and Mark O'Connell both have extensive experience as a result of many years of involvement in VCE General Mathematics (previously Further Mathematics). Each has taught the subject to hundreds of students of all abilities and, through this, has gained a valuable insight into the best approaches to use for each of the Areas of Study. Christine and Mark have been co-authors of many Further Mathematics trial papers, as well as acting in various writing, assessing and reviewing positions with the Victorian Curriculum and Assessment Authority (VCAA) and the Mathematical Association of Victoria (MAV).

9780170465335

UNIT 3

Chapter 1 Data analysis
Area of Study 1: Data analysis, probability and statistics

Content summary notes

Data analysis

Types of data

- Organising data
- Graphs
- Describing graphs
- Sample summary statistics for numerical data
- Measures of spread
- Boxplot (box-and-whisker plot)
- Investigating association between two variables

Fitting a linear model to a set of data

- Estimation of the equation of the least squares line
- Finding equation of a linear model drawn on a scatterplot
- Residual plots
- Transformations

Time series

- Investigating seasonal data

Area of Study 1 Outcome 1

These are the key knowledge required for this chapter, please note that not all will be examinable.

- types of data: categorical (nominal and ordinal) and numerical (discrete and continuous)

- frequency tables, bar charts including segmented bar charts, histograms, stem plots, dot plots, and their application in the context of displaying and describing distributions

- logarithmic (base 10) scales, and their purpose and application

- the five-number summary and boxplots (including the designation and display of possible outliers)

- mean $\bar{x}$ and sample standard deviation $s = \sqrt{\dfrac{\sum(x - \bar{x})^2}{n - 1}}$

- the normal model and the 68–95–99.7% rule, and standardised values (z-scores)

- response and explanatory variables

- two-way frequency tables, segmented bar charts, back-to-back stem plots, parallel boxplots, and scatterplots, and their application in the context of identifying and describing associations

- correlation coefficient, r, its interpretation, the issue of correlation and cause and effect

- coefficient of determination, its interpretation

- least squares line and its use in modelling linear associations

- data transformation and its purpose
- time series data and its analysis.

Although you should become familiar with all of the key skills, not all the skills are required for the exam. The key skills for Unit 3 Area of Study 1 are:

- construct frequency tables and bar charts and use them to describe and interpret the distributions of categorical variables
- answer statistical questions that require a knowledge of the distribution(s) of one or more categorical variables
- construct stem and dot plots, boxplots, histograms and appropriate summary statistics and use them to describe and interpret the distributions of numerical variables
- answer statistical questions that require a knowledge of the distribution(s) of one or more numerical variables
- solve problems using z-scores and the 68–95–99.7% rule
- construct two-way tables and use them to identify and describe associations between two categorical variables
- construct parallel boxplots and use them to identify and describe associations between a numerical variable and a categorical variable
- construct scatterplots and use them to identify and describe associations between two numerical variables
- calculate the correlation coefficient, r, and interpret it in the context of the data
- answer statistical questions that require a knowledge of the associations between pairs of variables
- determine the equation of the least squares line giving the coefficients correct to a required number of decimal places or significant figures as specified, and distinguish between correlation and causation
- use the least squares line of best fit to model and analyse the linear association between two numerical variables and interpret the model in the context of the association being modelled
- calculate the coefficient of determination, r^2, and interpret in the context of the association being modelled and use the model to make predictions, being aware of the problem of extrapolation
- construct a residual analysis to test the assumption of linearity and, in the case of clear non-linearity, transform the data to achieve linearity and repeat the modelling process using the transformed data
- identify key qualitative features of a time series plot including trend (using smoothing if necessary), seasonality, irregular fluctuations and outliers, and interpret these in the context of the data
- calculate, interpret and apply seasonal indices
- model linear trends using the least squares line of best fit, interpret the model in the context of the trend being modelled, use the model to make forecasts with consideration of the limitations of extending forecasts too far into the future.

VCE Mathematics Study Design 2023–2027 pp. 89–90, © VCAA 2022

1.1 Types of data

Categorical data

A category is recorded when the data is collected.

Examples of categorical **variables** include *gender, nationality, occupation.*

Categorical data can be further divided into nominal and ordinal data:

- **nominal data**: data that has no natural order, for example, *eye colour* (blue, brown, hazel)

- **ordinal data**: data that has a natural order, for example, *coffee cup size* (small, medium, large).

Numerical data

When data is collected, a number is counted or measured. There are two types of **numerical data**:

- **discrete data**: the numbers recorded are distinct values, often whole numbers, and usually come from counting; for example, *number of students in a class* or *pages in a book*

- **continuous data**: any number on a continuous number line is recorded; usually the data is produced by measuring to any desired level of accuracy; for example, *volume of water consumed*, or *the life of a battery.*

1.1.1 Organising data

Frequency tables may be used to organise both categorical and numerical data. Class intervals are used to group continuous numerical data or discrete data where there is a large range of values.

Stem-and-leaf plots (**stem plots**) display the distribution of numerical data (both discrete and continuous), as well as the actual data values. The median, interquartile range and range can be determined from an ordered stem plot.

Stem	Leaf
0	4 5 5
1	0 0 2 4 4 6 9
2	0 1 2 2 5 8
3	1 5

1.1.2 Graphs

Bar chart: used to display categorical data.

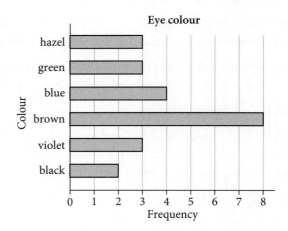

Percentage segmented bar chart: used to display categorical data.

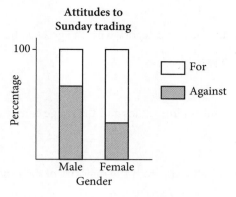

Column graph: used to display discrete data. Columns are of equal width with spaces in between. The heights of the columns represent the frequency.

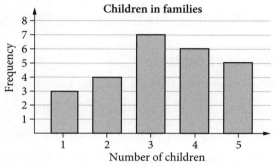

Histograms: used to display continuous data. Columns are joined together.

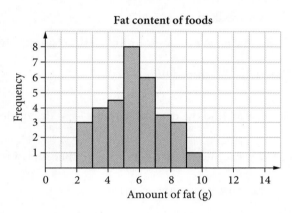

Dot plot: used to display discrete data.

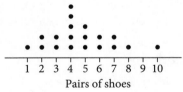

1.1.3 Describing graphs

Shape

Generally, one of three types:

- symmetric (same shape on either side of the centre)

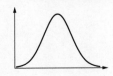

- positively skewed (tails off to the right)

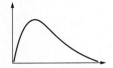

- negatively skewed (tails off to the left).

Centre

The mode (data value or category with the greatest frequency) or modal class interval (histogram) can be considered.

Spread

Maximum and minimum values and the range can be considered.

Outliers

Extreme values well away from the majority of data should be considered.

1.1.4 Sample summary statistics for numerical data

Measures of centre

Mean, $\bar{x}$: the average; the sum of all the data values divided by the number of values

Because all values are included in the calculation, the mean can be distorted by extreme values (**outliers**).

Median, Q_2: the middle value when the data is ordered

The median depends on position in the set of data and so is not distorted by extreme values and outliers.

To find the median:

1 Order the n data values from smallest to largest, where n is the number of data values.

2 Find the $\dfrac{n+1}{2}$ th position.

3 If n is odd, find the data value in the $\dfrac{n+1}{2}$ th position.

4 If n is even, find the two data values either side of the $\dfrac{n+1}{2}$ th position and average them.

Mode: the most frequently occurring value

A data set with two distinct modes is said to be bi-modal. The mode is not a relevant statistic in every set of data; it is often not representative of the data set or is not unique.

Measures of spread

Range = maximum value – minimum value

Interquartile range (IQR): the spread of the middle 50% of the data

IQR = upper quartile – lower quartile, i.e. IQR = $Q_3 - Q_1$.

Standard deviation, s: found using CAS. The formula for the standard deviation is $s = \sqrt{\dfrac{\Sigma(x - \bar{x})^2}{n-1}}$.

In comparisons of data sets, a larger standard deviation means the values are more spread out, or show more variation.

For bell-shaped (normal) distributions:

- 68% of the data will have values within one standard deviation of the mean

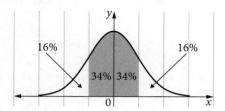

- 95% of the data will have values within two standard deviations of the mean

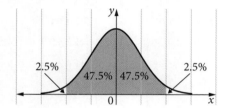

- 99.7% of the data will have values within three standard deviations of the mean.

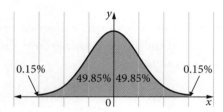

z-scores are used to compare values from different distributions. A **raw score**, x, from a data set will have a z-score of $\frac{x - \overline{x}}{s}$, where $\overline{x}$ is the mean of the data set and s is the standard deviation. The closer the z-score is to zero, the closer the raw score is to the mean. A positive z-score is greater than the mean; a negative z-score is less than the mean.

1.1.5 Boxplot (box-and-whisker plot)

A **boxplot** illustrates the **five-number summary** (minimum, lower quartile (Q_1), median (Q_2), upper quartile (Q_3), maximum) of a univariate numerical data set and is often used as a visual comparison of similar data sets. The median, IQR and range can readily be observed from a boxplot.

Outliers are identified using the lower fence: $Q_1 - 1.5 \times IQR$ and the upper fence: $Q_3 + 1.5 \times IQR$. Data values outside these fences are marked with a dot on the boxplot.

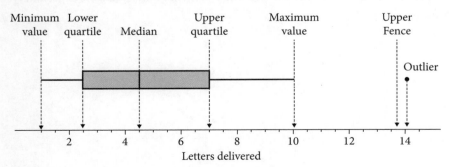

1.1.6 Investigating association between two variables

To identify **response** (dependent) and **explanatory** (independent) variables, ask the question: 'Which of the two variables is likely to change in response to a change in the other?' or 'Which of the two variables is likely to depend on the other?'

Back-to-back stem plots

Back-to-back stem plots are used to display a relationship between a numerical variable and a two-valued categorical variable. For example, comparing the *time* in minutes (numerical variable) of tram and train journeys (the alternatives of a two-valued categorical variable, *mode of transport*) to place of work.

Travel times - train	Stem	Travel times - tram
8 5 3 2	2	6 9
9 6 6 4 1 1 0	3	0 3 4 6 8
4 2 2 0 0	4	0 1 2 5 9 9
0	5	1 1 4 7
1 0	6	
	7	1

Key: 0 | 3 = 30   Key: 2 | 9 = 29

Parallel boxplots

Parallel boxplots are used to display a relationship between a numerical variable and a two-or-more-level categorical variable. For example, comparing the life expectancies from three different years (categorical variable *years* and numerical variable *life expectancy*).

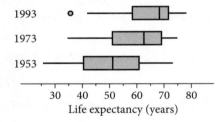

Tables (two-way frequency tables) and percentage segmented bar charts

Tables (two-way frequency tables) and percentage segmented bar charts are used to display the relationship between two categorical variables. For example, comparing male and female attitudes to Sunday trading (categorical variables: *gender* and *attitude*). An example of a percentage segmented bar chart is shown on page 5.

Scatterplots

Scatterplots are used to display the relationship between two numerical variables. For example, the relationship between *male life expectancy* (response variable) and *female life expectancy* (explanatory variable).

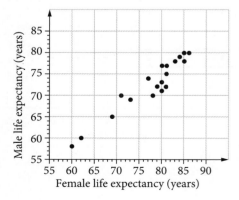

Description of scatterplots:

− direction: either positive, negative or random

− form: either linear or non-linear

− strength: either weak, moderate or strong

− outliers: any points that are separated from the general body of points.

Example on CAS

The data shows the *fat content*, in grams (explanatory variable), in list 1, and *energy*, in kilojoules (response variable), in list 2, of nine different foods.

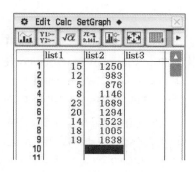

The scatterplot indicates a moderate, positive, linear relationship (correlation) between the variables *energy* and *fat content* with no outliers. As the *fat content* increases, the *energy* available in the food also increases.

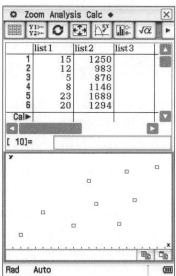

Pearson's correlation coefficient, r

r is a measure of the strength of the linear relationship between two variables. The value of r is between -1 (perfect negative linear relationship) and 1 (perfect positive linear relationship). If $r = 0$, then there is no correlation.

Assumptions made when using **Pearson's correlation coefficient**:

1 The variables are both numerical.

2 The association is linear.

3 There are no outliers.

Interpretation of r values

Ignoring the positive or negative sign:

• 0.75 to 1 is normally regarded as a strong correlation

• 0.5 to < 0.75 is normally regarded as a moderate correlation

• 0.25 to < 0.5 is normally regarded as a weak correlation

• 0 to < 0.25 is normally regarded as very little or no correlation.

Note
Correlation between two variables does not imply that a change in one variable causes a change in the other variable.

The correlation statistic screen for the *fat content* and *energy* is given on the right.

The *r* value is 0.69, confirming the observation that there is a moderate correlation between the variables.

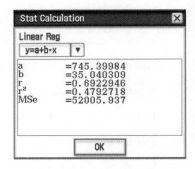

Coefficient of determination, r^2

It is the square of Pearson's correlation coefficient. The coefficient of determination is usually expressed as a decimal between 0 and 1. It can be converted to a percentage between 0 and 100%, which gives the percentage of the variation in the response variable that is explained by the explanatory variable.

For the variables *fat content* and *energy* from the previous example, the CAS screen shows an r^2 value of 0.4793 ($0.692\,32^2 \approx 0.4793$). This is interpreted as '47.93% of the variation in *energy* can be explained by the variation in *fat content*'.

> **Note**
> $0 \le r^2 \le 1$

1.2 Fitting a linear model to a set of data

Fitting a straight line (often called a linear model) to a scatterplot of bivariate numerical data is often called **linearisation**.

Two methods of fitting a linear model:

1 **Fitting a line 'by eye'**: using your judgement for the placement of the line. Used when the graph, but perhaps not the data, is available.

2 **The least squares regression line** is found using CAS (data must be available) and is a universally accepted method for applying a linear model. It should only be used when the graphed data appears to be linear and does not have any outliers.

> **Note**
> In the previous example, the **least squares line** has the equation
> *energy* = 745 + 35 × *fat content*

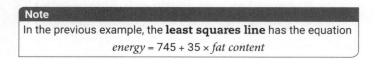

- The linear models can be used to predict either outside the given data range (**extrapolation**) or within the given data range (**interpolation**).

- Interpretation of the gradient or slope (35.04 in the previous CAS example):

  "On average, for each increase of one gram of fat, the energy increases by 35 kilojoules."

- Interpretation of intercept:

  'The energy in a food that contains zero grams of fat is 745 kilojoules.' If the intercept value is impractical, then the linear model is not suitable for predicting low values.

- residual value = actual value − predicted value

  A residual plot tests the quality of fit of a linear model. If the points on a residual plot are randomly scattered, then the linear model is appropriate for the data.

1.2.1 Estimation of the equation of the least squares line

- The equation is of the form $y = a + bx$, where $b = \dfrac{rs_y}{s_x}$ and $a = \bar{y} - b\bar{x}$.

- $\bar{x}$ and $\bar{y}$ are the means of x and y respectively.

  s_x and s_y are the standard deviations of x and y respectively.

- r is Pearson's correlation coefficient.

1.2.2 Finding the equation of a linear model drawn on a scatterplot

1 Choose two points on the line, (x_1, y_1) and (x_2, y_2).

2 Use $b = \dfrac{y_2 - y_1}{x_2 - x_1}$ for the gradient.

3 If the y-intercept (a) can be read directly from the line (if the x scale starts at zero), then use $y = a + bx$ for the equation. Otherwise, use the gradient and one point (x_1, y_1) in $y - y_1 = b(x - x_1)$ for the equation.

1.2.3 Residual plots

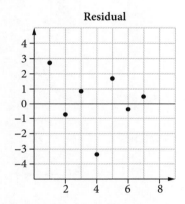

For the residual plot above, the linear modelling is appropriate as the points are scattered about the zero line.

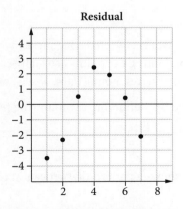

For the residual plot above, the linear model is **not** appropriate as there is a definite pattern to the points around the zero line.

1.2.4 Transformations

When a linear model is not appropriate, then a set of data can be transformed to linearity by applying a **transformation** to one of the axes. The transformation(s) that linearises the data will depend on the underlying shape of the points on the original scatterplot.

- x^2 stretches the x scale.

- $\log_{10}(x)$ compresses the x scale.

- $\dfrac{1}{x}$ compresses the large x values and stretches the small x values.

- y^2 stretches the y scale.

- $\log_{10}(y)$ compresses the y scale.

- $\dfrac{1}{y}$ compresses the large y values and stretches the small y values.

1.3 Time series

If time is the explanatory variable and data is collected over successive intervals of time, then the data is called a *time series*. A line graph joining the points is usually constructed for a time series. The successive intervals of time can be minutes, hours, days, weeks, months, years, etc.

Time series graphs can be described by identifying the following features:

- **trend**: upward or downward trend where the points are increasing or decreasing, respectively, as time progresses

- **seasonality**: the time series graph shows a pattern that repeats at regular intervals, in systematic, calendar-related movements

- **irregular fluctuations**: unsystematic, short-term fluctuations

- possible **outliers** and their sources: this includes one-off real-world events (e.g. war, pandemic) and signs of structural change such as a discontinuity in the time series.

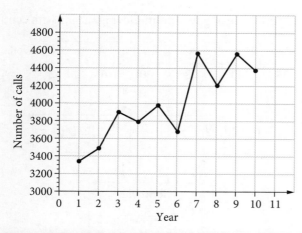

This time series graph displays an upward trend along with regular fluctuations.

> **Note**
>
> All time series graphs show some irregular fluctuations and more than one characteristic may be present. For example, a seasonal graph may also display a downward trend.

1.3.1 Investigating seasonal data

- **De-seasonalising data**

  A process that is used to remove the seasonal effects from a set of data. This allows any underlying trend to be made clearer.

- **Seasonal indices**

  A seasonal index is a measure of how a particular season compares with the average season. These are calculated during the process of de-seasonalisation.

$$\textbf{de-seasonalised figure} = \frac{\text{actual figure}}{\text{seasonal index}}$$

- Interpreting seasonal indices

  A seasonal index of $1.30 = \left(1 + \dfrac{30}{100}\right)$ for summer tells us that figures for summer are generally 30% higher than in an average season.

  A seasonal index of $0.94 = \left(1 - \dfrac{6}{100}\right)$ for June tells us that figures for June are generally 6% lower than in an average month.

  The average of all seasonal indices is always 1.

- **Smoothing**

  Smoothing removes irregular fluctuations from time series data. This allows any underlying trend to be seen more clearly. Techniques are **moving means smoothing** for numerical smoothing or **moving medians smoothing** for graphical smoothing (odd number of points only). Centring is required when the number of terms being smoothed is an even number.

  – **Moving means smoothing**

    Moving means smoothing on a graph is done by finding the mean of consecutive data points. The first and last data points do not have values when smoothed.

    For example:

Time period	1	2	3	4	5	6
Number of guests	35	43	78	80	63	52

  The smoothed value for time period 4, using three-point moving means smoothing, is

$$\frac{78 + 80 + 63}{3} = 73.67$$

To find the smoothed value for time period 4 using two-point moving means with centring:

Time period	1	2	3	4	5	6
Number of guests	35	43	78	80	63	52

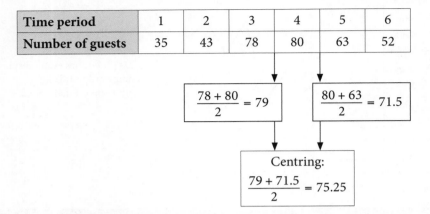

$$\frac{78 + 80}{2} = 79$$

$$\frac{80 + 63}{2} = 71.5$$

Centring:
$$\frac{79 + 71.5}{2} = 75.25$$

The number of points used when smoothing seasonal data should be the same as the number of seasons. For example, ice-cream sales over the four climatic seasons (summer, autumn, winter, spring) should be smoothed using a four-point moving means, then centred.

- **Moving medians smoothing**

Moving medians smoothing on a graph is done by progressively considering the points on the graph.

The example on the following page has been smoothed using three-point median smoothing.

Consider the first three points: $(1, 3)$, $(2, 6)$, $(3, 4)$. The median value on the horizontal axis $(1, 2, 3)$ is 2. The median value on the vertical axis $(3, 4, 6)$ is 4. So the median of the first three points is $(2, 4)$.

Consider the next three points: $(2, 6)$, $(3, 4)$, $(4, 10)$. The median value on the horizontal axis is 3 $(2, 3, 4)$. The median value on the vertical axis is 6 $(4, 6, 10)$. So, the median of the next three points is $(3, 6)$.

This process is repeated until all points are smoothed. The smoothed graph will always be shorter than the data graph by one point at either end.

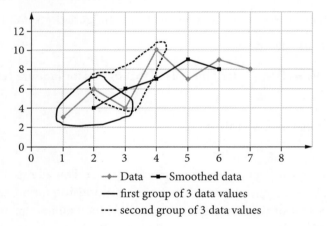

- **Trend lines**

A least-squares line can be fitted to time series data that appear linear.

- **Forecasting**

Making predictions of future values. Trend lines are often fitted after data has been smoothed or de-seasonalised and these models can be used for forecasting. Considerations are to be given to the possible limitations of fitting a linear model and the limitations of extending into the future.

- **Re-seasonalising data**

De-seasonalised forecasts can be 're-seasonalised' using

re-seasonalised prediction = de-seasonalised prediction × seasonal index

Glossary

back-to-back stem plots Used to display a relationship between a numerical variable and a two-valued categorical variable.

Travel times Route B	Stem	Travel times Route A
8 5 3 2	2	6 9
9 6 6 4 1 1 0	3	0 3 4 6 8
4 2 2 0 0	4	0 1 2 5 9 9
0	5	1 1 4 7
1 0	6	
	7	1

Key: 0 | 3 = 30   Key: 2 | 9 = 29

bar chart Used to display categorical data.

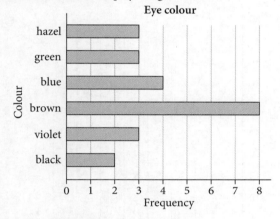

bell-shaped (normal) distribution

- 68% of the data will have values within one standard deviation from the mean

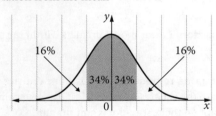

- 95% of the data will have values within two standard deviations from the mean

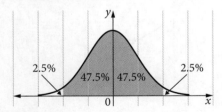

- 99.7% of the data will have values within three standard deviations from the mean

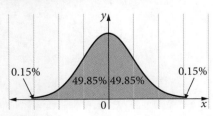

boxplot A boxplot (box-and-whisker plot) is used as a visual comparison of similar data sets. The statistics: median, IQR and range can readily be observed from a boxplot.

boxplot without outliers

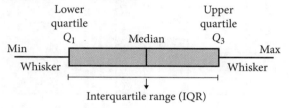

boxplot with outliers

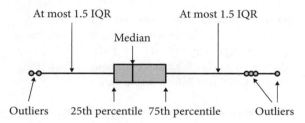

categorical data There are two types of categorical data: ordinal and nominal. *See also* **ordinal data**, **nominal data**.

coefficient of determination, r^2 The square of Pearson's correlation coefficient where $0 \leq r^2 \leq 1$.

A CAS screen shows both r and r^2.

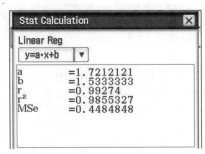

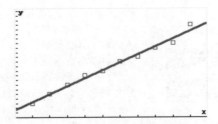

regression line: = $1.53 + 1.72x$, $r = 0.9927$, $r^2 = 0.9855$

column graph Used to display discrete data. Columns are of equal width with spaces between. The height of the columns represents the frequency.

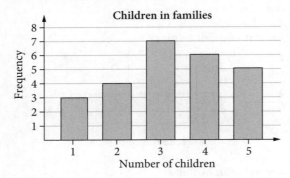

continuous data Any number on a continuous number line is recorded; usually the data is produced by measuring to any desired level of accuracy. There are two types of numerical data: continuous and discrete. *See also* **numerical data**, **discrete data**.

data Data are the specific values of a variable. A variable is something that can be counted, measured or categorised. *See also* **variables**.

de-seasonalising data A process that is used to remove the seasonal effects from a set of data.

$$\text{de-seasonalised figure} = \frac{\text{actual figure}}{\text{seasonal index}}$$

De-seasonalised forecasts can be 're-seasonalised' using the formula: re-seasonalised prediction = de-seasonalised prediction × seasonal index

discrete data Numerical variables that can only take specific values and can't be measured to ever increasing levels of accuracy. There are two types of numerical variables: continuous and discrete. *See also* **numerical data**, **continuous data**.

dot plot Used to display discrete data.

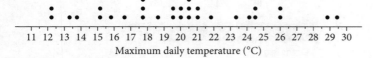

extrapolation Linear model predictions outside the given data range.

five-number summary
- minimum
- lower quartile: Q_1
- median: Q_2
- upper quartile: Q_3
- maximum

forecasting Making predictions of future values using trend lines.

frequency tables May be used to display both categorical and numerical data. Class intervals are used to group continuous numerical data or discrete data where there is a large range of values.

histogram Used to display continuous data; columns joined together.

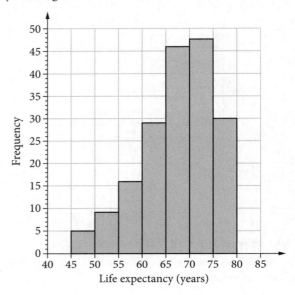

interpolation Linear model predicts within the given data range.

least squares regression line is a line drawn so that the sum of the squares of the vertical distance from each point on the scatterplot is a minimum.

- Fitting a line 'by eye':

9780170465335

- On CAS:

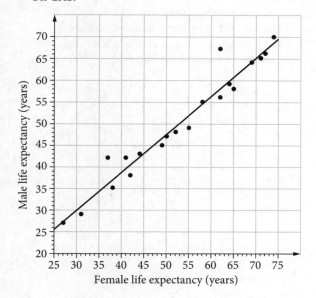

linearisation When an association of data is not linear, we apply a transformation to one of the variables so that the association between the two variables becomes closer to a straight line. This is called linearisation. *See also* **transformations**.

measures of centre

mean, $\bar{x}$: average

median, Q_2: the middle value when the data is ordered

mode: the most frequently occurring value

measures of spread

range = maximum value – minimum value

interquartile range, IQR: the middle 50% of the data

IQR = upper quartile – lower quartile, IQR = $Q_3 - Q_1$

standard deviation, s: using CAS

moving means smoothing A numerical smoothing technique that involves finding a series of means of a fixed number of data points.

moving medians smoothing A graphical smoothing technique that involves finding a series of medians of a fixed number of data points.

nominal data Categorical variables that have no natural/logical order. There are two types of categorical variables: ordinal and nominal. *See also* **categorical data, ordinal data**.

numerical data When data is collected, a number is counted or measured. There are two types of numerical variables: continuous and discrete. *See also* **continuous data, discrete data**.

ordinal data Categorical variables that have a natural/logical order. There are two types of categorical variables: ordinal and nominal. *See also* **categorical data, nominal data**.

organising data Visual displays of data include tables, plots and graphs.

outlier Any data values outside the fences found by the formulas:

Lower fence: $Q_1 - 1.5 \times IQR$

Upper fence: $Q_3 + 1.5 \times IQR$

parallel boxplots A statistical graph where two or more boxplots are shown on the same axis.

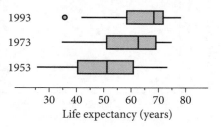

Pearson's correlation coefficient, *r* Measures the strength of the linear relationship between two variables, assuming linear data.

r falls between $-1 \leq r \leq 1$

$r = -1$ perfect negative linear relationship

$r = 1$ perfect positive linear relationship

Interpretation of r values, ignoring the positive or negative sign.

Pearson's correlation coefficient, *r*	Interpretation
0.75 to 1	strong correlation
0.5 to < 0.75	moderate correlation
0.25 to < 0.5	weak correlation
0 to < 0.25	very little or no correlation

percentage segmented bar chart Used to display categorical data.

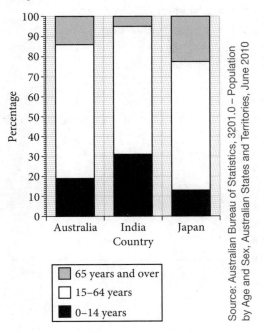

Source: Australian Bureau of Statistics, 3201.0 – Population by Age and Sex, Australian States and Territories, June 2010

raw score It is an actual data value.

scatterplot Scatterplots are used to display the relationship between two numerical variables. Used to demonstrate and visualise the relationship between two numerical variables.

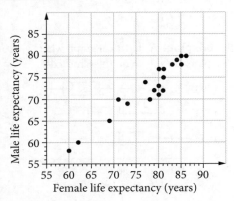

seasonal indices How a particular season compares with the average season, using de-seasonalisation.

smoothing Removes irregular fluctuations from time series data.

stem-and-leaf plots (stem plots) Display the distribution of numerical data (both discrete and continuous) as well as the actual data values.

Stem	Leaf
0	4 5 5
1	0 0 2 4 4 6 9
2	0 1 2 2 5 8
3	1 5

time series graphs A graph that shows numerical data with time as the independent value.

- Trend: upward or downward trend where the points are increasing or decreasing as time progresses.
- Seasonal: the time series graph shows a pattern that repeats at regular intervals.
- Irregular fluctuations: unsystematic, short-term fluctuations.
- Possible outliers and their sources

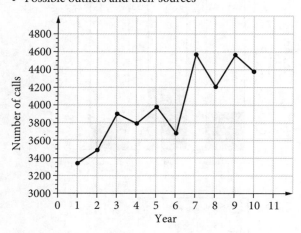

transformations When an association of data is not linear, we apply a transformation to one of the variables so that the association between the two variables becomes closer to a straight line. This is called linearisation. *See also* **linearisation**.

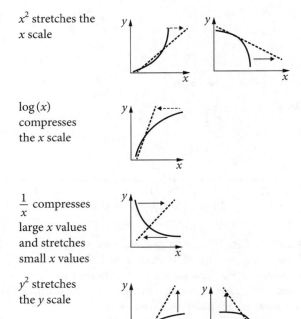

x^2 stretches the x scale

$\log(x)$ compresses the x scale

$\dfrac{1}{x}$ compresses large x values and stretches small x values

y^2 stretches the y scale

$\log(y)$ compresses the y scale

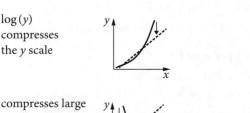

compresses large y values and stretches small y values

trend lines A least squares line can be fitted to time series data that appear linear.

types of data There are two main types of data: numerical and categorical.

variables A variable is something that can be counted, measured or categorised. Data are the specific values of a variable. *See also* **data**.

z-scores Used to compare values from different bell-shaped distributions. The raw score, x, has a z-score using the formula

$$z = \frac{x - \bar{x}}{s}$$

where $\bar{x}$ is the mean of the data set and s is the standard deviation. The closer the z-score is to zero, the closer the raw score is to the mean.

Exam practice

Multiple-choice questions

Investigating data distribution: 31 questions

Solutions to this section start on page 228.

Question 1

A single histogram was drawn to display the heights of 200 boys, comprising of 100 four-year-olds and 100 eight-year-olds who were randomly selected from different families. This distribution is most likely to be

A bi-modal

B negatively skewed

C positively skewed

D symmetrical

E segmented

Question 2

Which one of the following is **not** an example of discrete data?

A The number of pairs of school shoes purchased in a department store.

B The number of students in the class who have red hair.

C The total score obtained by the teams in an AFL match.

D The volume of cordial consumed by a group of primary students.

E The number of errors made on the multiple-choice section of a test.

Question 3

The distribution of the data on the dot plot given could best be described as

A symmetric

B positively skewed

C negatively skewed

D bell-shaped

E random

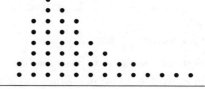

The following information relates to Questions 4 and 5.

The ordered stem plot below shows the weights, in kilograms, of a group of men.

5	2 3 1 5 5
6	0 2 3 5 6 6 7 8
7	4 4 4 6
8	3 7
9	1

Key: 5 | 2 means 52

Question 4

The number of men whose weights were recorded is

A 19   **B** 20   **C** 21   **D** 22   **E** 91

Question 5

The distribution of the weights could best be described as

A symmetrical

B bell-shaped

C positively skewed

D negatively skewed

E bi-modal

The following information relates to Questions 6 and 7.

The distribution of test marks obtained by a large group of students is displayed in the percentage frequency histogram below.

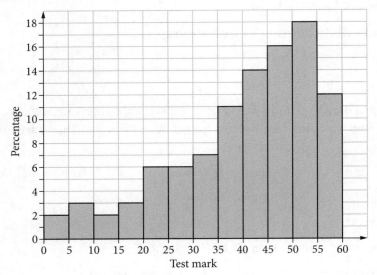

Question 6 ©VCAA 2006 1CQ5 ◐◑◑

The pass mark on the test was 30 marks. The percentage of students who passed the test is

A 7%   **B** 22%   **C** 50%   **D** 78%   **E** 87%

Question 7 ©VCAA 2006 1CQ6 ◐◑

The median mark lies between

A 35 and 40   **B** 40 and 45   **C** 45 and 50   **D** 50 and 55   **E** 55 and 60

Question 8 ◐◑

A set of discrete data has a median of 42, an upper quartile of 50 and a lower quartile of 32. To be identified as an outlier, a data value would have to be

A less than 15 or greater than 69.   **B** less than 24 or greater than 60.

C less than 5 or greater than 77.   **D** less than 17 or greater than 77.

E less than 14 or greater than 68.

The following information relates to Questions 9 and 10.

A set of data is displayed in the ordered stem-and-leaf plot below.

Stem	Leaf
0	4 5 5
1	0 0 2 4 4 6 9
2	0 1 2 2 5 8
3	1 5

Key: 1 | 2 means 12

Question 9 ◐◑

The median of this set of data is

A 16   **B** 17.5   **C** 18   **D** 19   **E** 19.5

Question 10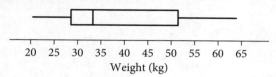

The interquartile range for this set of data is

A 10   **B** 12   **C** 15   **D** 18   **E** 22

Question 11

Consider the following boxplot representing the weights, in kilograms, of a large group of children.

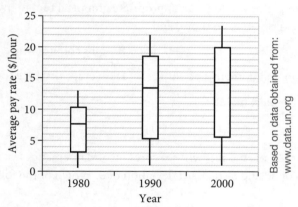

Which one of the following statements is true for this data?

A The interquartile range of weights is less than 20 kg.

B Less than a quarter of the weights are less than 32 kg.

C More than a quarter of the weights are greater than 49 kg.

D More than half the weights are less than 32 kg.

E The median weight is 35 kg.

Question 12 ©VCAA 2011 1CQ5

The boxplots below display the distribution of average pay rates, in dollars per hour, earned by workers in 35 countries for the years 1980, 1990 and 2000.

Based on the information contained in the boxplots, which one of the following statements is **not** true?

A In 1980, over 50% of the countries had an average pay rate less than $8.00 per hour.

B In 1990, over 75% of the countries had an average pay rate greater than $5.00 per hour.

C In 1990, the average pay rate in the top 50% of the countries was higher than the average pay rate for any of the countries in 1980.

D In 1990, over 50% of the countries had an average pay rate less than the median average pay rate in 2000.

E In 2000, over 75% of the countries had an average pay rate greater than the median average pay rate in 1980.

Question 13 🞉

The given frequency table represents the number of students absent in Year 12 at Ellis High over a 20-day period.

Absentees	0	1	2	3	4	5	6
Frequency	3	2	5	3	0	5	2

The median number of absentees during these 20 days is

A 2   **B** 2.5   **C** 2.9   **D** 3   **E** 5

Question 14 🞉

When preparing a boxplot of his students' results on a test, a teacher noticed that one of the group of 20 was an outlier. It would be **true** to say that this student

A had not studied for the test.

B scored more than the upper quartile plus 1.5 × the interquartile range.

C scored less than the lower quartile minus 1.5 × the interquartile range.

D was either the highest or lowest scorer on this test.

E had a score that was at least 2 standard deviations away from the mean.

Question 15 ©VCAA 2014 1CQ8 🞉

A single back-to-back stem plot would be an appropriate graphical tool to investigate the association between a car's speed, in kilometres per hour, and the

A driver's age, in years.   **B** car's colour (white, red, grey, other).

C car's fuel consumption, in kilometres per litre.   **D** average distance travelled, in kilometres.

E gender of the driver (female, male).

Question 16 🞉

Mark is part of a large group of students whose scores on a Mathematics test were found to be normally distributed with a mean of 60 and a standard deviation of 6. Which of the following scores is Mark most likely to have achieved?

A 42   **B** 48   **C** 55   **D** 68   **E** 72

Question 17 🞉

Ivan has completed 15 multiple-choice tests and has counted the number of errors made in each.

These are displayed in the following frequency table.

Errors	0	1	2	3	4
Frequency	1	4	6	1	3

Which one of the following statements is **true** regarding the number of errors?

A The mean is greater than the median.   **B** The median is greater than the mean.

C The mean is equal to the median.   **D** The distribution is bimodal.

E The median is greater than the mode.

Question 18 ●●

A General Mathematics class did two tests, one on univariate data and the other on bivariate data. It was found that every student in the class achieved 2 marks more on the bivariate data test than they did on the univariate data test. In comparison with the mean and standard deviation of the univariate data test, the mean and standard deviation of the bivariate data test will be respectively

A unchanged and unchanged.

B 2 greater and 2 greater.

C 2 greater and unchanged.

D unchanged and 2 greater.

E 2 greater and 2 less.

Question 19 ©VCAA 2005 1CQ6 ●●

The distribution of fuel consumption of a particular model of car is approximately bell-shaped with a mean of 8.8 km per litre and a standard deviation of 2.2 km per litre. The percentage of this model of car that has a fuel consumption less than 6.6 km per litre is closest to

A 2.5%   **B** 5%   **C** 16%   **D** 32%   **E** 68%

The following information relates to Questions 20 and 21.

When a sample of 200 large eggs are weighed, the distribution of weights is found to be bell-shaped with a mean of 65 g and a standard deviation of 1.2 g.

Question 20 ●●

The number of eggs in the sample that would be expected to have a weight between 63.8 g and 66.2 g would be

A 34   **B** 68   **C** 95   **D** 136   **E** 190

Question 21 ●●

An egg in the sample weighs 66.8 g. The standardised value (*z*-score) for this egg is

A 1   **B** 1.5   **C** 2   **D** 2.5   **E** 3

Question 22 ●●

The results of a statistics exam, when graphed, were found to produce the following bell-shaped curve.

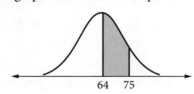

If the mean is 64 and the standard deviation is 11, then the percentage of students whose scores are represented by the shaded region is closest to

A 11%   **B** 22%   **C** 34%   **D** 47.5%   **E** 50%

The following information relates to Questions 23 and 24.

The heights of twenty-five Year 10 students are measured and recorded to the nearest centimetre.

The results are displayed on the split-stem plot below.

Stem	Leaf (Height)
15	4
15*	6 6
16	0 3 4 4
16*	5 6 6 6 7 7 9
17	0 0 1 3 3 4
17*	5 6 6
18	0 2
18*	

Key: 15 | 4 means 154
15* | 6 means 156

Question 23 ●●○

The percentage of students with a height of at least 170 cm is

A 11%   **B** 36%   **C** 40%   **D** 44%   **E** 48%

Question 24 ●●○

The interquartile range for this set of data is

A [164, 174]   **B** [164, 173]   **C** 9   **D** 9.5   **E** 10

Question 25 ●●○

Amanda scored 73 on a test where the class mean was 60 and the standard deviation was 6.3.

Amanda's standardised score (z-score), correct to two decimal places, is

A −1.11   **B** 1.11   **C** 2.06   **D** 13.00   **E** 63.48

Question 26 ●●○

In an exam, the mean score was 64 and the standard deviation was 10. An exam mark corresponding to a z-score of −0.7 is

A 55   **B** 56   **C** 57   **D** 63.3   **E** 71

Question 27 ©VCAA 2015 1CQ5 ●●○

The foot lengths of a sample of 2400 women were approximately normally distributed with a mean of 23.8 cm and a standard deviation of 1.2 cm.

The standardised foot length of one of these women is $z = -1.3$. Her actual foot length, in centimetres, is closest to

A 22.2   **B** 22.7   **C** 25.3   **D** 25.6   **E** 31.2

Question 28 ●●●

400 young women were weighed and the results were found to be normally distributed with a mean of 56 kilograms and a standard deviation of 8 kilograms. The number of these women who have a weight between 48 kilograms and 72 kilograms is expected to be

A 190   **B** 200   **C** 272   **D** 326   **E** 380

Question 29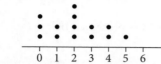

A sample of 14 people were asked to indicate the time (in hours) they had spent watching television the previous night. The results are displayed in the dot plot below.

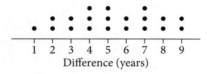

Time spent watching television (hours)

Correct to one decimal place, the mean and standard deviation of these times are respectively

A $x = 2.0\ s = 1.5$

B $x = 2.1\ s = 1.5$

C $x = 2.1\ s = 1.6$

D $x = 2.6\ s = 1.2$

E $x = 2.6\ s = 1.3$

Question 30 ⬤⬤⬤

The number of cars in each household of all the houses in a street are counted and the results are displayed in the following frequency table.

Number of cars	Frequency
0	1
1	7
2	10
3	4
4	2
5	0

For this set of data, the mean and standard deviation, respectively, of the number of cars per household, correct to two decimal places, are

A 2 and 1.00

B 2.17 and 1.09

C 2.04 and 0.93

D 1.96 and 1.00

E 2.04 and 0.91

Question 31 ©VCAA 2015 1CQ3 ⬤⬤

The dot plot below displays the difference between female and male life expectancy, in years, for a sample of 20 countries.

Difference (years)

The mean ($\overline{x}$) and standard deviation (s) for this data are

A mean = 2.32, standard deviation = 5.25

B mean = 2.38, standard deviation = 5.25

C mean = 5.0, standard deviation = 2.0

D mean = 5.25, standard deviation = 2.32

E mean = 5.25, standard deviation = 2.38

Investigating associations between two variables: 48 questions

Solutions to this section start on page 230.

Question 32

A scatterplot involving two variables y and x is presented below.

The value of Pearson's correlation coefficient would be closest to

A −0.6   **B** −0.2   **C** 0.2   **D** 0.6   **E** 1

Use the following information to answer Questions 33 and 34.

The blood pressure (low, normal, high) and the age (under 50 years, 50 years or over) of 110 adults were recorded. The results are displayed in the two-way frequency table below.

Blood pressure	Age	
	Under 50 years	**50 years or over**
low	15	5
normal	32	24
high	11	23
Total	58	52

Question 33 ©VCAA 2016 1CQ1

The **percentage** of adults under 50 years of age who have high blood pressure is closest to

A 11%   **B** 19%   **C** 26%   **D** 44%   **E** 58%

Question 34 ©VCAA 2016 1CQ2

The variables of *blood pressure* (low, normal, high) and *age* (under 50 years, 50 years or over) are

A both nominal variables.

B both ordinal variables.

C a nominal variable and an ordinal variable respectively.

D an ordinal variable and a nominal variable respectively.

E a continuous variable and an ordinal variable respectively.

The following information relates to Questions 35 and 36.

The following two-way frequency table shows the party preference of 100 randomly selected male and female American voters who were surveyed prior to a presidential election.

Party preference	Gender		Total
	Female	Male	
Democrat	30	24	54
Republican	33	13	46
Total	63	37	100

Question 35

The percentage males surveyed is

A 13%   **B** 24%   **C** 37%   **D** 46%   **E** 54%

Question 36

The percentage of females who intended to vote for the Republicans is closest to

A 30%   **B** 33%   **C** 35%   **D** 52%   **E** 72%

Question 37

Which of the following would be the most appropriate means of displaying the performances of three classes of students on a Geography test?

A two-way frequency table   **B** histogram   **C** segmented bar chart

D back-to-back stem plot   **E** parallel boxplots

Question 38

Pearson's correlation coefficient for the scatterplot shown is closest to

A −1

C −0.2

E 0.7

B −0.7

D 0.5

Question 39

The following graph shows two boxplots representing a class' performance on Test *A* and on Test *B*.

In comparison to Test *A*, the results on Test *B* are

A generally higher and more variable.

B generally higher and less variable.

C generally lower and more variable.

D generally lower and less variable.

E largely unchanged.

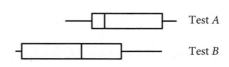

Test *A*

Test *B*

CHAPTER 1 – EXAM PRACTICE

Use the following information to answer Questions 40 and 41.

In New Zealand, rivers flow into either the Pacific Ocean (the Pacific rivers) or the Tasman Sea (the Tasman rivers).

The boxplots below can be used to compare the distribution of the lengths of the Pacific rivers and the Tasman rivers.

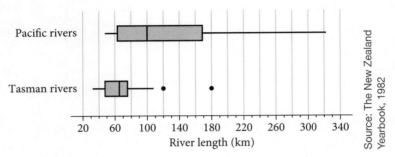

Source: The New Zealand Yearbook, 1982

Question 40 ©VCAA 2015 1CQ6 ●●

The five-number summary for the lengths of the Tasman rivers is closest to

A 32, 48, 64, 76, 108

B 32, 48, 64, 76, 180

C 32, 48, 64, 76, 322

D 48, 64, 97, 169, 180

E 48, 64, 97, 169, 322

Question 41 ©VCAA 2015 1CQ7 ●●

Which one of the following statements is **not** true?

A The lengths of two of the Tasman rivers are outliers.

B The median length of the Pacific rivers is greater than the length of more than 75% of the Tasman rivers.

C The Pacific rivers are more variable in length than the Tasman rivers.

D More than half of the Pacific rivers are less than 100 km in length.

E More than half of the Tasman rivers are greater than 60 km in length.

Question 42 ●●

A scatterplot of exam performance versus hours of preparation for a group of students is shown below.

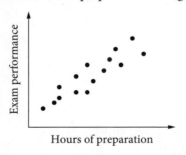

The value of the coefficient of determination for these variables would be closest to

A 0   **B** 0.3   **C** 0.7   **D** 0.95   **E** 1

The following information relates to Questions 43, 44 and 45.

The level of mobile phone usage (never, use family phone, owner) for a sample of 196 secondary students is indicated in the table below. Some of the entries in the table are missing.

Level of mobile phone usage	Age			
	12–13	14–15	16–17	Total
Never	12	14	10	
Use family phone		28		75
Phone owner		21	40	
Total			60	196

Question 43 ⬤⬤⬜

For this sample, the total number of students who are phone owners is

A 24   **B** 61   **C** 85   **D** 111   **E** 135

Question 44 ⬤⬤⬜

The variables *age* and *level of phone usage* for this two-way frequency table are respectively, examples of

A categorical and numerical variables.   **B** nominal and ordinal variables.

C numerical and ordinal variables.   **D** ordinal and nominal variables.

E numerical and categorical variables.

Question 45 ⬤⬤⬜

The percentage of 12–13-year-old students who use the family phone is closest to

A 19%   **B** 33%   **C** 37%   **D** 42%   **E** 51%

Question 46 ⬤⬤⬜

A study of the relationship between stress levels and work productivity involving 100 factory workers concluded that an increase in stress tended to be associated with a decrease in productivity.

The coefficient of determination was calculated to be 0.65. The value of Pearson's correlation coefficient, correct to two decimal places, is

A −0.81   **B** −0.42   **C** 0.42   **D** 0.65   **E** 0.81

Question 47 ⬤⬤⬜

Which one of the following statements is **true** regarding Pearson's correlation coefficient, *r*?

A The value of *r* is always between 0 and 1.

B The squared value of *r* is a measure of the association between variables.

C A value for *r* of 1 is a stronger correlation than a value of −1.

D A high value of *r* means that one variable must be influencing the other.

E A value of 0 for *r* means that there is no correlation at all between the variables.

Question 48 ©VCAA 2016 1CQ8 ●●

Parallel boxplots would be an appropriate graphical tool to investigate the association between the monthly median rainfall, in millimetres, and the

A monthly median wind speed, in kilometres per hour.

B monthly median temperature, in degrees Celsius.

C month of the year (January, February, March, etc.).

D monthly sunshine time, in hours.

E annual rainfall, in millimetres.

Question 49 ●●

Jordan is investigating the relationship between gender and reading ability (5 levels) in a group of nine-year-old children. Which one of the following would be **true** for this investigation?

A Reading ability is the explanatory variable.

B Both variables are categorical variables.

C Histograms could be used to display the investigation results.

D Back-to-back stem plots could be used to display the investigation results.

E Parallel boxplots could be used to display the investigation results.

Question 50 ●●

A scatterplot involving two variables y and x is shown below.

If the correlation coefficient is 0.95, which of the following statements is the most accurate regarding the two variables?

A A change in y is caused by a change in x.

B A change in y tends to be associated with a change in x.

C 95% of the variation in y can be explained by the variation in x.

D The relationship between y and x is non-linear.

E No conclusion can be drawn as there are no scales on the axes.

Question 51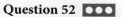

The parallel boxplots show the distributions of marks for three classes. There were 24 students in each class and they all sat for the same test.

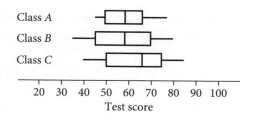

Which one of the following would **not** be true for the results?

A The highest mark was obtained by a student in Class C.

B The median for Class A is the same as the median for Class B.

C The range of results is the same for classes B and C.

D The number of students in Class C who scored 65 or more is double the number of students in Class B who scored 65 or more.

E At least 25% of all the students obtained a score of 50 or less.

Question 52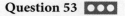

Which one of the following would **not** be true for the data displayed for the scatterplot shown?

A x is the explanatory variable in this case.

B If the data value marked (a) was removed, then the value of Pearson's correlation coefficient would increase to a value closer to 0.

C As the value of x increases, the value of y appears to decrease.

D There is a moderate negative association between x and y.

E For this set of data, the value of the coefficient of determination would be greater than the value of the Pearson's correlation coefficient.

Question 53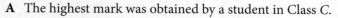

The association between the depth of tread on a car tyre and the number of kilometres travelled by a car was investigated. If 84% of the variation in depth of tread can be explained by the variation in the number of kilometres travelled by the car, which one of the following scatterplots would show the correlation between depth of tread and the number of kilometres travelled?

A

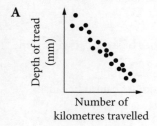

B

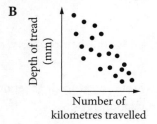

C

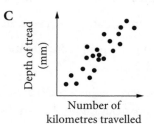

D

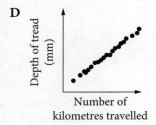

E

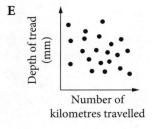

The following information relates to Questions 54 and 55.

The travel times between two towns along two different routes have been recorded and the times, in minutes, are given in the back-to-back stem plot below.

Travel times (Route B)	Stem	Travel times (Route A)
8 5 3 2	2	6 9
9 6 6 4 1 1 0	3	0 3 4 6 8
4 2 2 0 0	4	0 1 2 5 9 9
0	5	1 1 4 7
1 0	6	
	7	1

Key: 0 | 3 = 30 minutes   2 | 6 = 26 minutes

Question 54 ⬤⬤

The percentage of travel times, on either route, that were more than 60 minutes is closest to

A 5%   **B** 6%   **C** 8%   **D** 10%   **E** 12%

Question 55 ⬤⬤⬤

Which one of the following would be true for the data displayed in this stem plot?

A The distribution of the data for both routes is symmetric.

B The data value 71 minutes on Route *A* would be an outlier.

C The range is the same for both data sets.

D The interquartile range is larger for Route *B* than for Route *A*.

E There is a difference of 5.5 minutes in the median travel times.

Question 56   ©VCAA 2017 1CQ12   ⬤⬤

Data collected over a period of 10 years indicated a strong, positive association between the number of stray cats and the number of stray dogs reported each year ($r = 0.87$) in a large, regional city.

A positive association was also found between the population of the city and both the number of stray cats ($r = 0.61$) and the number of stray dogs ($r = 0.72$).

During the time that the data was collected, the population of the city grew from 34 564 to 51 055.

From this information, we can conclude that

A if cat owners paid more attention to keeping dogs off their property, the number of stray cats reported would decrease.

B the association between the number of stray cats and stray dogs reported cannot be causal because only a correlation of +1 or −1 shows causal relationships.

C there is no logical explanation for the association between the number of stray cats and stray dogs reported in the city so it must be a chance occurrence.

D because larger populations tend to have both a larger number of stray cats and dogs, the association between the number of stray cats and the number of stray dogs can be explained by a common response to a third variable, which is the increasing population size of the city.

E more stray cats were reported because people are no longer as careful about keeping their cats properly contained on their property as they were in the past.

The following information relates to Questions 57 and 58.

A survey is conducted to investigate an association between arthritis (none, moderate, debilitating) and age (4 categories). The results are displayed on the following percentage segmented bar chart.

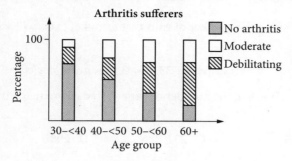

Question 57 ●●●

The percentage of people in the 50–<60 age group who have debilitating arthritis is approximately

A 15%   **B** 20%   **C** 25%   **D** 35%   **E** 50%

Question 58 ●●●

Which one of the following would **not** be a conclusion that could be made from this survey?

A As age increases, there appears to be an increase in the percentage of people who are affected by arthritis.

B Approximately 75% of the people in the 60+ age group are affected by arthritis.

C Exactly twice as many people in the 60+ age group as people in the 30–<40 age group are affected by arthritis.

D More than 50% of those surveyed in the 50–<60 age group were affected by arthritis.

E 50% of those in the 30–<40 group who have no arthritis can expect to be affected to some degree by the time they reach 60.

Question 59 ●●●

The age and fitness level of a group of industrial workers is recorded and the results are shown on the scatterplot. One outlier is shown in this set of data.

If the outlier in this group is removed, then the value of Pearson's correlation coefficient will

A increase to a value closer to 1.

B decrease to a value closer to 0.

C increase to a value closer to –1.

D decrease to a value closer to –1.

E not change.

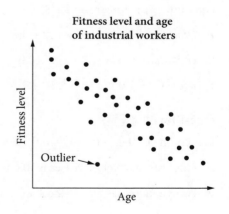

Investigating and modelling linear associations: 20 questions

Solutions to this section start on page 232.

Question 60 ⬤▢▢

A least squares line for predicting the number of runs that a cricket team will score from the number of overs bowled is as follows.

$$runs\ scored = -33.42 + 3.53 \times overs\ bowled$$

On a day when 75 overs were bowled, the number of runs scored as predicted by this model will be closest to

A 42 **B** 105 **C** 232 **D** 298 **E** 300

The following information relates to Questions 61 and 62.

Josh is investigating the association between age and cholesterol level and his data is plotted on the scatterplot below. He has fitted a linear model 'by eye' to the scatterplot.

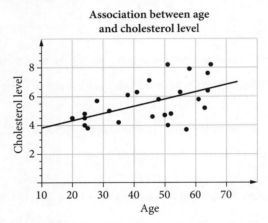

Question 61 ⬤⬤▢

Josh has found the slope of this line to be 0.05. If the line goes through the point (40, 5.3), then the equation of the linear model is

A *cholesterol level* = 0.05 × *age* + 3.7 **B** *cholesterol level* = 0.05 × *age* + 3.3

C *cholesterol level* = 3.7 × *age* + 0.05 **D** *age* = 0.05 × *cholesterol level* + 3.3

E *age* = 0.05 × *cholesterol level* + 3.7

Question 62 ⬤⬤▢

Which one of the following would **not** be a true interpretation of Josh's linear model?

A Cholesterol level increases by 0.05 units per year.

B As age increases, cholesterol level increases.

C An increase in age causes an increase in cholesterol level.

D The line would be overpredicting for the person aged 50 whose cholesterol level is 4.7.

E Using this line to predict the cholesterol level of a person aged 70 would be a case of extrapolation.

The following information relates to Questions 63 and 64.

The average rainfall and temperature range at several different locations in the South Pacific region are displayed in the scatterplot below.

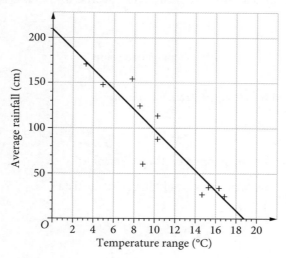

Question 63 ©VCAA 2004 1CQ8 ●●

A least squares regression line has been fitted to the data as shown.

The equation of this line is closest to

A *average rainfall* = 210 − 11 × *temperature range*.

B *average rainfall* = 210 + 11 × *temperature range*.

C *average rainfall* = 18 − 0.08 × *temperature range*.

D *average rainfall* = 18 + 0.08 × *temperature range*.

E *average rainfall* = 250 − 13 × *temperature range*.

Question 64 ©VCAA 2004 1CQ9 ●●

The value of the Pearson's correlation coefficient, r, for the data, is $r = -0.9260$.

The value of the coefficient of determination is

A −0.9260   **B** −0.8575   **C** 0.8575   **D** 0.9260   **E** 0.9623

Question 65 ©VCAA 2014 1CQ9 ●●

The equation of a least squares regression line is used to predict the fuel consumption, in kilometres per litre of fuel, from a car's weight, in kilograms. This equation predicts that a car weighing 900 kg will travel 10.7 km per litre of fuel, while a car weighing 1700 kg will travel 6.7 km per litre of fuel.

The slope of this least squares regression line is closest to

A −250   **B** −0.005   **C** −0.004   **D** 0.005   **E** 200

Question 66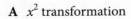

The following scatterplot of y versus x is clearly non-linear.

Which of the following transformation(s) can be used to linearise the data?

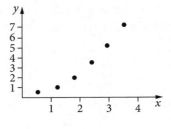

A x^2 transformation

B y^2 transformation

C $\dfrac{1}{x}$ transformation

D $\log_{10}(x)$ transformation

E $\dfrac{1}{x}$ transformation followed by a $\log_{10}(x)$ transformation

Question 67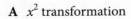

Two numerical variables, x and y are graphed on this scatterplot.

Which one of the following transformations would linearise the data?

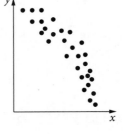

A $\log_{10}(x)$

B x^2

C $\log_{10}(y)$

D $\dfrac{1}{x}$

E $\dfrac{1}{y}$

Question 68   ©VCAA   2000 1CQ9

A least squares line is the line for which

A as many data points lie on the line as possible.

B the sum of the squares of the vertical distances from the line to each data point is a minimum.

C the sum of the squares of the horizontal distances from the line to each data point is a minimum.

D the sum of the squares of the shortest distances from the line to each point is a minimum.

E the data set is divided so that as many points are above the line as below the line.

Question 69

The least squares line equation for two variables, Q and P, is found to be $P = 82.5 - 8.6Q$.

The residual for the data point $(3.5, 46.4)$ is

A −52.4   **B** −46.4   **C** −6   **D** 6   **E** 52.4

Question 70

Data has been collected to investigate the association between *energy supplied* and *fat content* for a number of brands of muesli.

Percentage of fat	Energy (kilojoules) per 100 g
2.4	1236
3.6	1330
7.8	1845
15.4	2067
8.2	1683
5.7	1520

The value of Pearson's correlation coefficient for this data, correct to two decimal places, is

A 0.86   **B** 0.89   **C** 0.90   **D** 0.95   **E** 0.96

Question 71 ⬤⬤⬤

A $\log_{10}$ transformation is applied to the explanatory variable of a set of bivariate numerical data.
Which one of the following sets of data will be linearised by this transformation?

A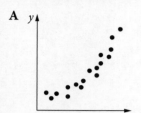

B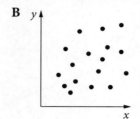

C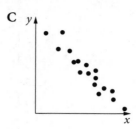

D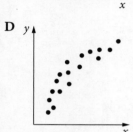

E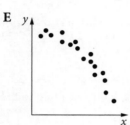

Question 72 ⬤⬤⬤

A least squares line has been applied to the data graphed on the scatterplot as shown.

Which one of the following would be the residual plot for this scatterplot?

A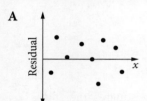

B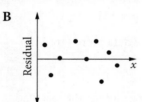

C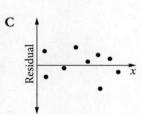

D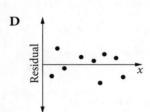

E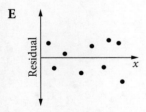

The following information relates to Questions 73 and 74.

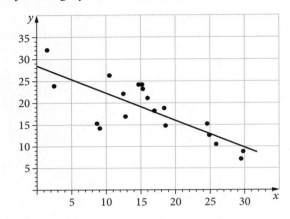

Question 73 ●●●

A least squares line with equation $y = 28.5 - 0.62x$ has been fitted to the scatterplot above.

Which one of the following is **not** true for this line?

A It is the line for which the sum of the squares of the vertical distances from the line to each data point is a minimum.

B The slope can be interpreted as 'for each one unit increase in x, y increases by 0.62 units'.

C Using the line to predict y for $x = 20$ is a case of interpolation.

D The residual for the point $(17, 18.2)$ will be positive.

E The line underpredicts the value of y for the data point $(16, 21.1)$.

Question 74 ●●●

A point $(5, 10)$ has been omitted from the scatterplot for the calculation of the least squares line. If this point is included, which one of the following will be the effect on the least squares line?

A The slope will increase in value, the intercept will decrease.

B The slope will decrease in value, the intercept will decrease.

C The slope and the intercept will remain the same.

D The slope will increase in value, the intercept will increase.

E The slope will decrease in value, the intercept will increase.

Question 75 ©VCAA 2009 1CQ12 ●●

The *mathematics achievement* level (TIMSS score) for grade 8 students and the general rate of *internet use* (%) for 10 countries are displayed in the scatterplot.

To linearise the data, it would be best to plot

A *mathematics achievement* against *internet use*.

B log (*mathematics achievement*) against *internet use*.

C *mathematics achievement* against log (*internet use*).

D *mathematics achievement* against (*internet use*)².

E $\dfrac{1}{mathematics\ achievement}$ against *internet use*.

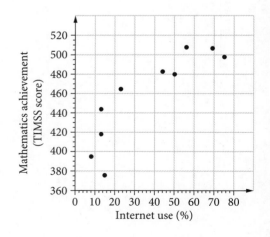

Question 76 ●●●

A $\log_{10}$ transformation, applied to either the explanatory or response variable, is to be used to linearise a set of bivariate numerical data. Which one of the following data sets will be linearised by a log transformation?

A

B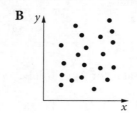

C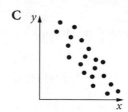

D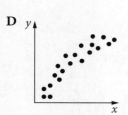

E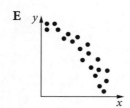

Question 77 ●●●

A least squares line for a set of data has the equation

number of fish caught = 0.6 + 2.2 × *number of hours spent fishing*

On one particular day, Dave spent 2.4 hours fishing and he caught two fish. Using the least squares line, the residual for this data point would be closest to

A −5.9   **B** −3.9   **C** −3.3   **D** 3.3   **E** 5.3

The following information relates to Questions 78 and 79.

The height and weight of a sample of women are measured. The height of the group has a mean of 165.5 cm and a standard deviation of 5.2 cm. The weight of the group has a mean of 62 kilograms and a standard deviation of 6.8 kilograms. The correlation between the weight and height of the women is investigated and Pearson's correlation coefficient, r, between their weight and height is found to be 0.86. A least squares line is fitted to this data in order to predict weight from height.

Question 78 ●●●

The slope of the line, correct to two decimal places, will be

A 0.37   **B** 0.66   **C** 1.12   **D** 1.31   **E** 41.12

Question 79 ●●●

The intercept of the least squares line, to the nearest kilogram, will be

A −380   **B** −154   **C** −124   **D** −65   **E** 1

Investigating and modelling time series data: 21 questions

Solutions to this section start on page 233.

Question 80 ●●●

Which one of the following is a feature of all time series graphs?

A seasonality   **B** irregular fluctuations   **C** one-off real world events

D increasing trend   **E** decreasing trend

Question 81 ◔◌◌

The number of wedding receptions per month in a large hotel has been recorded in the table below.

Month	Jan	Feb	Mar	Apr	May	Jun	Jul	Aug
Number of receptions	15	12	19	22	28	30	23	19

Using three-point moving means, the smoothed value for April will be

A $17\frac{2}{3}$ **B** 22   **C** 23   **D** $28\frac{2}{3}$ **E** $15\frac{1}{3}$

Question 82 ◕◌◌

The time series on the right could best be described as

A irregular

B seasonal

C seasonal with trend

D trend only

E repetitive

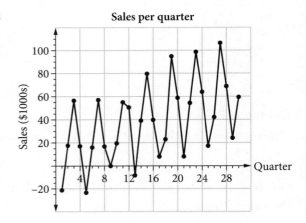

Sales per quarter

Question 83 ◕◌◌

The following table shows seasonal indices at Santino's café for this year.

Quarter 1	Quarter 2	Quarter 3	Quarter 4
1.00	0.79	1.09	

The missing seasonal index for Quarter 4 will be equal to

A 0.70   **B** 0.96   **C** 1.00   **D** 1.12   **E** 1.21

Question 84 ◕◌◌

The number of sedans sold per month by a major distributor is given in the table below.

Month	Jan	Feb	Mar	Apr	May	Jun	Jul	Aug
Number of sedans sold	101	122	137	118	112	104	97	88

The data is to be smoothed using four-point moving means with centring. The smoothed value for April will be

A 113.63   **B** 114   **C** 117.5   **D** 120   **E** 121.25

The following information relates to Questions 85 and 86.

The table below gives the seasonal indices for the four quarters of a year and the de-seasonalised data for the years 2016 and 2017.

Quarter	1	2	3	4
Seasonal index	0.935	1.124	1.086	0.855
De-seasonalised sales for 2016 ($1000s)	1002	1056	1127	1158
De-seasonalised sales for 2017 ($1000s)	1176	1247	1283	1319

A least-squares line based on the de-seasonalised data with equation

$$de\text{-}seasonalised\ sales = 44.43 \times time\ period + 971$$

has been fitted to the de-seasonalised data using the 1st quarter of 2016 as time period 1, 2nd quarter of 2016 as time period 2 … 4th quarter of 2017 as time period 8.

Question 85

Using the model, the predicted de-seasonalised value for the 3rd quarter of 2018 will be

A 1104   **B** 1282   **C** 1439   **D** 1460   **E** 1586

Question 86

The re-seasonalised prediction of sales, in thousands of dollars, for the 3rd quarter of 2018 will be closest to

A 1017   **B** 1199   **C** 1344   **D** 1563   **E** 1586

Question 87

The seasonal index for the June quarter 2016 of a seasonal time series data is calculated to be 0.836.

Which one of the following statements will **not** be true for this data?

A The sales for the June quarter were less than the average quarterly sales for 2016.

B The sales in one or more of the other quarters would have been greater than those for the June quarter.

C The sales for the June quarter were 16.4% less than the average quarterly figure for 2016.

D To de-seasonalise the sales for June quarter, we would multiply by 0.836.

E The seasonal index for at least one of the other quarters in 2016 would be greater than one.

Question 88 ©VCAA 2015 1CQ12

The time series plot on the right charts the number of calls per year to a computer help centre over a 10-year period.

Using five-median smoothing, the smoothed number of calls in year 6 was closest to

A 3500   **B** 3700

C 3800   **D** 4000

E 4200

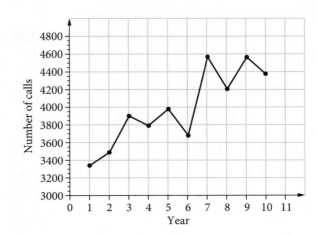

Question 89 ●●●

The given table shows the number of days above 30°C in Whiteville for the first eight months of 2017.

Month	Jan	Feb	Mar	Apr	May	Jun	Jul	Aug
No. of days	14	12	9	7	4	5	5	8

Four-point moving means smoothing with centring is to be applied to this data. The smoothed value for the month of March will be

A 9   B 9.25   C 9.5   D 9.75   E 10

Question 90 ●●●

Some data is recorded in the table below.

Time period	1	2	3	4	5	6
Data	12	14	16	21	17	15

The data above is to be smoothed using both three-point median smoothing and three-point moving means smoothing.

The smoothed values using three-point median and moving means, respectively, for time period 4 will be

A 21 and 18   B 18 and 16   C 17 and 17

D 21 and 17   E 17 and 18

Use the following information to answer Questions 91, 92 and 93.

The table below shows the long-term average of the number of meals served each day at a restaurant. Also shown is the daily seasonal index for Monday through to Friday.

	Day of the week						
	Mon	Tues	Wed	Thurs	Fri	Sat	Sun
Long-term average	89	93	110	132	145	190	160
Seasonal index	0.68	0.71	0.84	1.01	1.10		

Question 91 ©VCAA 2016 1CQ14 ●●

The seasonal index for Wednesday is 0.84.

This tells us that, on average, the number of meals served on a Wednesday is

A 16% less than the daily average.   B 84% less than the daily average.

C the same as the daily average.   D 16% more than the daily average.

E 84% more than the daily average.

Question 92 ©VCAA 2016 1CQ15 ●●

Last Tuesday, 108 meals were served in the restaurant.

The de-seasonalised number of meals served last Tuesday was closest to

A 93   B 100   C 110   D 131   E 152

Question 93 ©VCAA 2016 1CQ16 ●●●

The seasonal index for Saturday is closest to

A 1.22   **B** 1.31   **C** 1.38   **D** 1.45   **E** 1.49

Question 94 ●●●

The following table gives the quarterly data for the year 2018 and the seasonal indices for three of the seasons.

Quarter	1	2	3	4
Data for 2018	1424	1765	1828	1536
Seasonal index	0.84	1.14	1.09	

The seasonal index for the fourth quarter and the de-seasonalised data for the second quarter, 2018, are respectively

A 0.94 and 1878   **B** 1.02 and 1730   **C** 0.93 and 2012

D 1.02 and 1800   **E** 0.93 and 1548

Question 95 ●●●

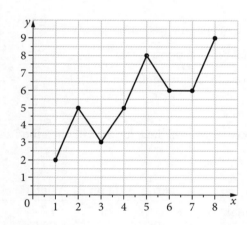

Using three-point median smoothing, the smoothed graph for the time series would be

A
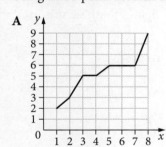

B
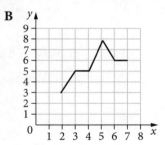

C
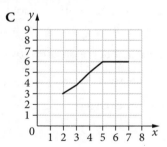

D
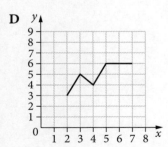

E
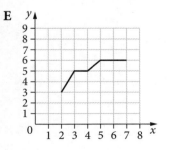

Question 96 ●●

The following graph shows the sales of laptops plotted for the first 10 months of operation of a shop.

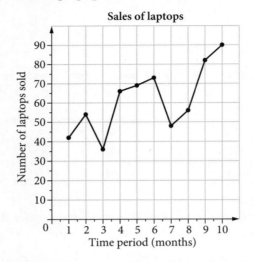

From this graph it can be said that laptop sales for the first 10 months showed

A seasonality with irregular fluctuations.

B irregular fluctuations only.

C an increasing trend only with irregular fluctuations.

D a decreasing trend with irregular fluctuations.

E an increasing trend with seasonality and irregular fluctuations.

Question 97 ●●●

The seasonal indices, based on several years of data, for the sales in a food shop have been calculated and are shown in the table below.

Season	Summer	Autumn	Winter	Spring
Seasonal index	0.88	0.96	1.11	1.05

Comparing actual sales figures for Autumn and Winter to de-seasonalised sales figures for these seasons, the de-seasonalised sales figures will

A increase and decrease respectively.   B decrease and increase respectively.

C both increase.   D both decrease.

E both be unchanged.

Question 98

The 2019 quarterly figures for the number of nights of accommodation booked in a popular resort town are given in the table below.

Quarter	1	2	3	4
Number of nights	2460	3643	1832	1415

Based on the figures for 2019 only, the seasonal index for the 3rd quarter is

A 0.221   **B** 0.503   **C** 0.724   **D** 0.784   **E** 1.276

Question 99 ©VCAA 2017N 1CQ14

The sales figures used to generate a time series plot are displayed in the table below.

Year	Quarter 1	Quarter 2	Quarter 3	Quarter 4
2013	6.5	13.4	7.4	3.8
2014	10.2	11.8	7.4	4.5
2015	9.6	14.5	8.6	5.3
2016	10.3	14.2	7.5	4.9

The four-mean smoothed sales with centring for Quarter 3 in 2015, in millions of dollars, was closest to

A 8.6   **B** 9.3   **C** 9.5   **D** 9.6   **E** 9.7

Question 100

The time series plot below shows the share price of two companies over a period of 20 months.

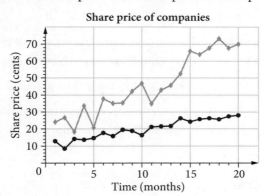

From the plot it can be concluded that over the interval of the 20 months, the difference in share price between the two companies has shown

A an increasing trend.   **B** a decreasing trend.   **C** seasonal variation.

D calendar-related movements.   **E** no trend.

Extended-answer questions

Solutions to this section start on page 235.

101 Approaching normality (15 marks)

Question 101.1 (11 marks)

Lennard is a General Mathematics teacher from a large school who has decided to undertake an analysis of the normal distribution. He knows that VCE General Mathematics study scores, when graphed, form a normal curve with a mean of 30 and a standard deviation of 7. He also knows that, practically, the highest and lowest study scores awarded are 50 and 10 respectively. Suppose that 20 000 students are doing General Mathematics in this particular year.

a The graph below represents the distribution of marks for the 20 000 students. How many students would expect their mark to be within the shaded region?   1 mark

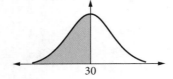

b Fill in the missing values in the given table.   2 marks

Score range	Percentage of students	Number of students
23–37		
16–44		

c Determine a possible range of marks that contains about 81.5% of students would lie.   2 marks

d Harry received a study score of 44 for General Mathematics this year. Write down the formula that would enable his standardised value (z-score) to be determined.   1 mark

e Determine the z-score for Harry and explain what it means in a practical context.   2 marks

Lennard looks back at Harry's marks on two earlier pieces of assessment during the year. These results are summarised in the table below. Both sets of results were found to follow the shape of the normal distribution.

Task	Mean	Standard deviation	Harry's result
SAC on Networks	55	18	46
SAC on Matrices	77	20	61

f Using z-scores, determine in which of the two tasks Harry was ranked higher in his class.   3 marks

Question 101.2 (4 marks)

a For the SAC on Recursion and Financial Modelling, the results for Harry's class were again normally distributed. If the middle 68% of scores were found to be between 36 and 64, determine the mean and standard deviation for this data.   2 marks

b Harry's classmate Jan received her SAC back but couldn't read whether the result was 20 or 30. Briefly explain which result would be more likely for Jan.   1 mark

c Comment on how likely it is that any student in Harry's class gained a perfect score of 100.   1 mark

102 Card club (15 marks)

A card club was formed and the attendance over the first nine weeks was recorded and graphed below.

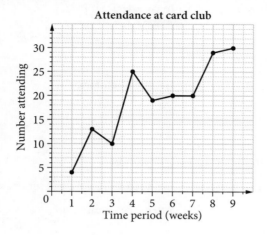

Attendance at card club

Question 102.1 (3 marks)

a   Comment on this time series graph.   1 mark

b   On the graph above, smooth the data using three-point median smoothing.   2 marks

Question 102.2 (5 marks)

In an attempt to find a better smoothed graph, it is decided to use four-point moving means smoothing with centring. The table below shows the smoothed values, correct to two decimal places, except for the value for time-period 6.

a   Calculate the value for time period 6.   2 marks

Time period	1	2	3	4	5	6	7	8	9
Attendance	4	13	10	25	19	20	20	29	30
Smoothed data			14.88	17.63	19.75		23.38		

b   Plot the smoothed points on the set of axes below.   2 marks

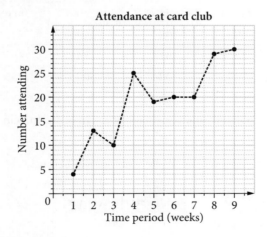

Attendance at card club

c   Comment on the smoothed graph.   1 mark

Question 102.3 (7 marks)

a Fit a least squares line to the smoothed data from Question 102.2, writing your values for the slope and intercept of this line, correct to two decimal places.

attendance = ☐ + ☐ × *time period*   2 marks

b Interpret the slope of this least squares line.   1 mark

c Use your least squares line equation to predict the attendance in

   i week 2   1 mark

   ii week 8   1 mark

   Give your answers correct to the nearest whole number.

d Use your results from part **c** to graph the least-squares line on the set of axes below.   2 marks

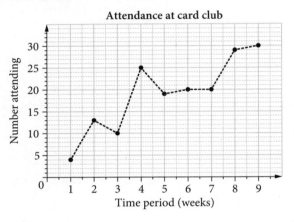

Attendance at card club

(y-axis: Number attending; x-axis: Time period (weeks))

103 Carmelo's gelati (15 marks)

Question 103 (15 marks)

Carmelo sells gelati and other refreshments from a small shop in his suburb. Sales of gelati are traditionally seasonal, with sales expected to be higher in the warmer weather. Carmelo has noted the number of gelati sold per quarter over a three-year period are as follows.

	Quarter 1	Quarter 2	Quarter 3	Quarter 4
2016	502	343	293	422
2017	487	314	254	465
2018	565	384	300	435

Carmelo needs your assistance to determine seasonal indices in order to help with future planning.

a Determine the quarterly average for 2018. Write your answer below.   1 mark

Year	Quarterly average
2016	390
2017	380
2018	

b Now express each quarter's figures for 2018 as a proportion of the quarterly average
for that year. Give each answer correct to three decimal places.   1 mark

	Quarter 1	Quarter 2	Quarter 3	Quarter 4
2016	1.287	0.879	0.751	1.082
2017	1.282	0.826	0.668	1.224
2018				

c To determine the seasonal index for each quarter, calculate the average of each column
in part **b** above. Write your answers, correct to two decimal places.   2 marks

	Seasonal index
Quarter 1	1.30
Quarter 2	
Quarter 3	
Quarter 4	

d Give an interpretation of the seasonal index for Quarter 1.   1 mark

e If Carmelo had used monthly sales instead of quarterly, what would be the total of the
seasonal indices?   1 mark

f Using the seasonal indices found in part **c**, complete the following table of de-seasonalised
sales figures. Give your answers correct to the nearest whole number.   2 marks

	Quarter 1	Quarter 2	Quarter 3	Quarter 4
2016	386	394		380
2017	375		358	419
2018	435	441	423	

g Using Quarter 1 2016 as time period 1, Quarter 2 2016 as time period 2, etc., determine the
equation for a least squares line in the form

$$de\text{-}seasonalised\ sales = a \times time\ period + b$$

Give *a* and *b* correct to two decimal places.   2 marks

h Use the equation to predict the de-seasonalised sales figures for Quarter 2 2018.
Give your answer correct to the nearest whole number.   1 mark

i Hence, calculate the residual value of de-seasonalised sales for Quarter 2 2018.   1 mark

j Use the equation found in part **g**, together with the seasonal indices to forecast the
sales for Quarter 4 2019. Give your final answer correct to the nearest whole number.   2 marks

k Is the forecast from part **j** an example of interpolation or extrapolation?   1 mark

104 Fatten-up (14 marks)

Two large fast-food chains offer a variety of individual meals. The fat content has been recorded for each of these meals.

Chain A fat content (g)	Chain B fat content (g)
15	12
12	13
32	13
22	22
23	19
18	20
8	24
5	23
27	14
	18
	12

Question 104.1 (3 marks)

A boxplot for the fat content for meals from chain *A* has been drawn below.

a Construct a boxplot for the fat content of meals from chain *B* using the same set of axes.   1 mark

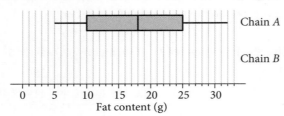

b Comment and compare the distributions of the data from each of the chains, referring to shape, centre and spread.   2 marks

Question 104.2 (6 marks)

The fat content, in grams, and the amount of energy provided, in kilojoules, was recorded for meals from food chain *B* and a scatterplot was drawn.

Fat content (g)	Energy (kilojoules)
12	601
13	660
13	681
22	1181
19	1012
20	1001
24	1216
23	1250
14	840
18	951
12	856

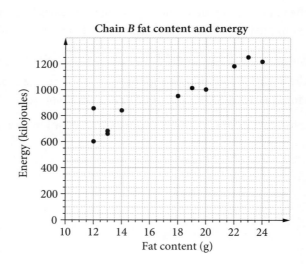

a Comment on the scatterplot in terms of direction, form and strength.   2 marks

b Find Pearson's correlation coefficient for the data and interpret the result.   2 marks

c Find the coefficient of determination for this data and interpret the result.   2 marks

Question 104.3 (5 marks)

a Fit a least squares line to the data and write the equation in terms
of the variables.   2 marks

b Use your equation to predict the energy provided by a meal that contains 30 g of fat.
Give your answer to the nearest kilojoule.   1 mark

c Using your least squares line as the model, find the residual for the data point (20, 1001),
giving your answer to the nearest kilojoule.   2 marks

105 Geoff and Helen (15 marks)

Question 105.1 (9 marks)

Geoff is a teacher who has a General Mathematics class of 16 students this year. The first test
of the year was out of 100 marks. The results were as follows.

63	47	40	32	56	65	56
59	62	90	47	59	65	50

a Calculate the mean and standard deviation for these results. Write your answers correct
to one decimal place.   2 marks

At Geoff's school there is a second General Mathematics class taught by Helen. The results
of Helen's class of 16 students on the same test are as follows.

52	36	99	75	86	45	90	71
54	53	80	47	59	72	70	21

b Calculate the mean and standard deviation for these results. Write your answers correct
to one decimal place.   2 marks

c Geoff and Helen decide to analyse the results with the aid of parallel boxplots. With
reference to the types of variables involved, explain why parallel boxplots are appropriate.   1 mark

The parallel boxplots are illustrated on the following graph.

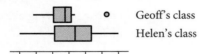

Geoff's class
Helen's class

d Compare the two boxplots in terms of shape, centre, spread and outliers.   4 marks

Question 105.2 (2 marks)

On the second test of the year, Geoff noticed that each member of his class scored two marks
higher than their result on the first test.

Describe the change from Test 1 to Test 2

a for the mean   1 mark

b for the standard deviation.   1 mark

Question 105.3 (4 marks)

Helen decides to investigate whether her class' results for Test 1 fit the properties of a normal (bell-shaped) distribution.

a   Determine the percentage of the class that are:

   i   within one standard deviation from the mean   1 mark

   ii   within two standard deviations from the mean   1 mark

   iii   within three standard deviations from the mean.   1 mark

b   Comment on whether this class' results approximate to a normal distribution.   1 mark

106 Global warming (15 marks)

Question 106 (15 marks)

The maximum and minimum temperature forecasts for 15 cities around the world on a particular day are given in the following table.

City	Maximum temperature	Minimum temperature
Athens	25	19
Bangkok	30	25
Beijing	22	19
Berlin	20	11
Cairo	31	24
Frankfurt	23	7
London	25	11
Los Angeles	19	18
Madrid	31	25
Moscow	16	8
New Delhi	32	25
Paris	21	11
Rome	26	16
Tokyo	34	24
Toronto	24	20

a   Explain why a scatterplot would be appropriate as a way of displaying this set of data.   1 mark

b   Most of the data has been graphed as a scatterplot onto the following axes. Complete the graph by plotting the data for Paris, Rome, Tokyo and Toronto.   1 mark

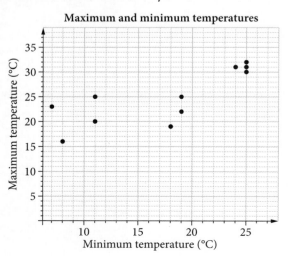

c From the scatterplot, describe the association between the two variables.   1 mark

d A least squares regression line is to be used to predict maximum temperature from
minimum temperature. Explain why a least squares regression line can be used with
this data.   1 mark

e Using CAS, determine the equation of the least squares regression line.   2 marks

f Draw the least squares line on the axes above. Label the coordinates of two points.   2 marks

g Which two cities are best predicted by the least squares line?   1 mark

h A residual plot is to be used to assess the suitability of the least-squares line. The table gives
the observed maximum temperatures, the maximum temperatures predicted by the line
(correct to one decimal place) and the residual. Fill in the missing values.   4 marks

City	Actual maximum temperature	Predicted maximum temperature	Residual
Athens	25	26.2	−1.2
Bangkok	30	30.1	−0.1
Beijing	22		−4.2
Berlin	20	21.0	−1.0
Cairo	31	29.5	1.5
Frankfurt	23	18.4	4.6
London	25	21.0	
Los Angeles	19	25.6	−6.6
Madrid	31	30.1	0.9
Moscow	16	19.1	−3.1
New Delhi	32		
Paris	21	21.1	−0.1
Rome	26	24.3	1.7
Tokyo	34	29.5	4.5
Toronto	24	26.9	−2.9

i Complete the residual plot for the minimum temperatures below by plotting the points for
Paris, Rome, Tokyo and Toronto.   1 mark

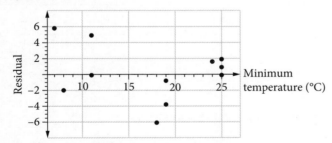

j What does the residual plot tell us about the suitability of the linear model?   1 mark

107 One out of the box (15 marks)

Question 107.1 (8 marks)

Two mathematics classes at KF College, each with 19 students, completed the same examination. Class *A* had the following percentage scores.

88	62	71	26	85	68	77	52	93	58
59	55	48	33	80	85	71	83	31	

a Using CAS, determine the mean and standard deviation correct to two decimal places.   2 marks

b Complete the following five-number summary for Class *A*.   1 mark

minimum value	
lower quartile	
median	
upper quartile	
maximum value	

c Construct a boxplot for the data of Class *A*.   1 mark

d Give a reason why the median is greater than the mean.   1 mark

e The teacher of Class *A*, Ray Rover, claims that, of the four sections of the boxplot, the section from the upper quartile to the maximum value contains the least number of students. Is Rover correct?   1 mark

f The data for Class *A* contained no outliers. Complete the following steps to prove this statistically.   2 marks

interquartile range (IQR) = ☐

$1.5 \times$ IQR = ☐

fences = $[Q_1 - 1.5 \times$ IQR$, Q_3 + 1.5 \times$ IQR$]$

= ☐

Conclusion _____

_____

_____

Question 107.2 (7 marks)

The data for Class *B* is given as follows, but one score is missing.

57	32	39	88	53	61	68	75	51
79	87	61	91	87	48	77	51	61

a Calculate the missing score from Class *B*, given that the two classes had identical means.   1 mark

b On the same axis as 107.1 part **c**, draw a boxplot of the data from Class *B* that enables the two classes to be compared.   1 mark

c Write a brief statement comparing the two sets of data with respect to

   i shape

   ii centre

   iii spread.   3 marks

d One of the students in Class *B*, Rod Rocket, scored the lowest mark of 32%.
By what margin did Rocket avoid being considered an outlier?   2 marks

108 **Linear behaviour** (15 marks)

Question 108.1 (10 marks)

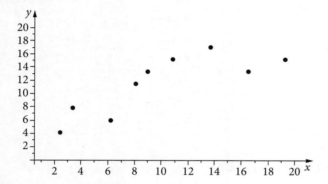

a Describe the scatterplot in terms of direction, form and strength.   3 marks

b Give an estimate of Pearson's correlation coefficient. What assumption are we making
in using this coefficient?   2 marks

c On the graph, draw a line of best fit, 'by eye'.   1 mark

d The equation of your 'line-of-best-fit' can be written in the form $y = a + bx$.
Using your line, determine approximate values for a and b.   2 marks

e Write a brief sentence explaining the significance of the constant a.   1 mark

f Write a brief sentence explaining the significance of the constant b.   1 mark

Question 108.2 (5 marks)

Some statistics have been provided to allow for an equation for the least squares line that will
predict y from x. The values given are the means of x and y, the standard deviations of x and y
and Pearson's correlation coefficient.

$$\bar{x} = 9.5 \qquad \bar{y} = 11.6 \qquad s_x = 5.75 \qquad s_y = 4.5 \qquad r = 0.8$$

a Determine the slope of the least squares line, correct to two decimal places.   1 mark

b Determine the intercept of the least squares line, correct to two decimal places.   1 mark

c Use the equation to predict the value of y when x equals 7.6, correct to two decimal places.   1 mark

d If the actual value of y is 12 when $x = 7.6$, determine the value of the residual for this
prediction from part **c**.   1 mark

e If the slope of the least squares line remained the same, determine what the new intercept
must be in order for the value of the residual to be 0.   1 mark

109 Share and share alike (15 marks)

Question 109.1 (8 marks)

The price of shares can fluctuate considerably, but on one particular day all the following shares increased in value.

Company	Increase in value of 1 share (dollars)
ABC	0.15
DEF	0.54
GHI	0.20
JKL	0.28
MN	0.20
OP	0.06
QR	1.93
STU	0.08
VW	0.02
XYZ	0.54

a Determine the median value for this set of data.   1 mark

b Determine the mean for this set of data.   1 mark

c Give a reason why the median is lower than the mean.   1 mark

d Complete the following five-number summary.   2 marks

minimum	0.02
lower quartile	
median	
upper quartile	
maximum	1.93

e The increase in value of a QR share of $1.93 is an outlier amongst this data. Showing all your calculations, determine how much lower this value needs to be in order to not be considered as an outlier.   3 marks

Question 109.2 (3 marks)

The actual value of each share at the close of trading on this day for each company is as follows.

Company	Value of 1 share (dollars)
ABC	7.32
DEF	19.14
GHI	9.55
JKL	36.23
MN	21.91
OP	3.63
QR	47.28
STU	17.32
VW	5.26
XYZ	15.56

a Using the information in both tables, determine the percentage increase for Woolworths shares from the start of the day to the finish. Give your answer correct to one decimal place.   2 marks

b If the percentage increase of a XYZ's share the following day was 5%, determine the new share value at the end of the day.   1 mark

Question 109.3 (2 marks)

All purchasers of VW shares 'on-line' on this particular day were asked to provide some demographic information. In analysing the income of each person, it was found that this data was approximately normally distributed with a mean of $120 000 and a standard deviation of $30 000.

Determine the percentage of VW share purchasers who had an income

a between $90 000 and $150 000   1 mark

b less than $180 000.   1 mark

Question 109.4 (2 marks)

If the income for OP share purchasers was bell-shaped with a mean of $140 000 and approximately 84% had an income less than $160 000, determine the standard deviation for this data.   2 marks

110 Tough brake (15 marks)

The distance travelled after the brakes are first applied until a car comes to a complete stop is recorded for cars travelling at different speeds.

Speed (km/h)	25	45	60	70	80	90	100
Stopping distance (m)	45	51	62	85	105	118	144

A scatterplot of the data is shown. A least squares regression line has been applied to the data as shown below.

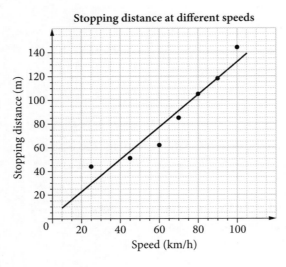

Stopping distance at different speeds

Question 110.1 (3 marks)

a Sketch a residual plot for the data.   2 marks

b Give a reason why the linear model is not a reasonable fit for the data.   1 mark

Question 110.2 (7 marks)

Observing that the change in stopping distance with the increase in speed is clearly not linear, it is decided to apply a square transformation to the explanatory variable.

a Complete the following table.   1 mark

Speed (km/h)	25	45	60	70	80	90	100
(Speed)^2							
Stopping distance (m)	44	51	62	85	105	118	144

b Complete the scatterplot of the transformed data by plotting the last three points from the table on the set of axes below.   2 marks

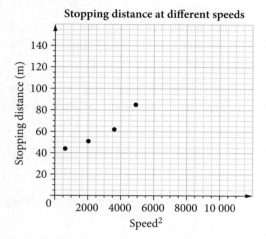

c Find the equation of the least squares line for the transformed data. Write the coefficients, correct to three decimal places, in the spaces provided.

stopping distance = ⬜ + ⬜ × *speed*^{2} 2 marks

d Use the least-squares equation to predict the stopping distance for a car that is travelling at 110 km/h when the brakes are applied. Give your answer correct to one decimal place.   2 marks

Question 110.3 (5 marks)

The residual plot that results from fitting a least-squares line to this transformed data is shown below. The point that corresponds with a speed value of 100 km/h is missing.

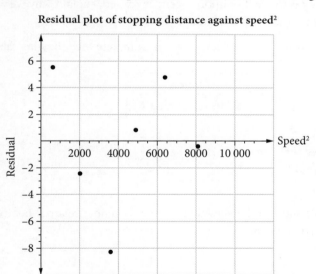

a Find the residual that would correspond with a speed of 100 km/h and plot this value on the graph.   4 marks

b This residual plot suggests that the stopping distance versus speed^2 model is a good fit to the data. Which feature of this residual plot supports this conclusion?   1 mark

111 Two-way (15 marks)

Question 111.1 (9 marks)

One hundred people were randomly selected and surveyed as to whether they were in favour of 4-year terms for the position of Premier of Victoria. The group consisted of 58 males and 42 females and each person voted 'for' or 'against'.

When the data was tabulated, it was found that from the male group, 30 voted for the proposal and the remainder were against, whereas from the female group, 29 voted for the proposal and the remainder were against.

a Which of the two variables, *opinion* or *gender*, is the explanatory variable?   1 mark

b Complete the following sentence.

A two-way frequency table is appropriate here, as both *opinion* (for or against) and *gender* (male or female) are [   ] variables.   1 mark

c Complete the two-way frequency table.   2 marks

<table>
<tr><td rowspan="2" colspan="2"></td><td colspan="3">Gender</td></tr>
<tr><td>Male</td><td>Female</td><td>Total</td></tr>
<tr><td rowspan="3">Opinion</td><td>For</td><td></td><td></td><td></td></tr>
<tr><td>Against</td><td></td><td></td><td></td></tr>
<tr><td>Total</td><td>58</td><td>42</td><td>100</td></tr>
</table>

d If the intention of this survey is to see if opinion is gender related, explain why we would use column percentages rather than row percentages.   1 mark

e Using column percentages, correct to one decimal place, complete the following table.   2 marks

<table>
<tr><td rowspan="2" colspan="2"></td><td colspan="2">Gender</td></tr>
<tr><td>Male</td><td>Female</td></tr>
<tr><td rowspan="3">Opinion</td><td>For</td><td></td><td></td></tr>
<tr><td>Against</td><td></td><td></td></tr>
<tr><td>Total</td><td>100%</td><td>100%</td></tr>
</table>

f Do the results from the table support the view that opinion is gender related? Justify your answer by quoting appropriate percentages.   2 marks

Question 111.2 (6 marks)

From the group surveyed, 15 men and 15 women were asked to give their age. The data, in raw form, is as follows.

Men   31, 45, 76, 21, 62, 72, 18, 23, 60, 51, 49, 37, 51, 48, 55
Women   24, 19, 71, 28, 21, 33, 46, 29, 39, 65, 51, 57, 41, 38, 22

a Illustrate this data using a back-to-back stem plot.   2 marks

Leaf (Men)	Stem (Age)	Leaf (Women)

b Compare the two distributions, quoting statistics where appropriate, with regard to

  i shape   1 mark

  ii centre   1 mark

  iii spread   1 mark

  iv outliers.   1 mark

112 Vital statistics (15 marks)

Question 112 (15 marks)

The heights and weights of a group of AFL footballers were measured as follows.

Height (centimetres)	Weight (kilograms)
181	85
186	91
202	95
190	92
195	95
189	94
182	93
192	86
178	75

a For this data, use CAS to determine Pearson's correlation coefficient, r, correct to four decimal places.   1 mark

b Comment briefly on what the value of r tells us about the strength of the association between height and weight.   1 mark

c State the coefficient of determination, correct to four decimal places.   1 mark

d Interpret the coefficient of determination in terms of the variables *weight* and *height*.   1 mark

e What are the lowest and highest values that the coefficient of determination may take?   2 marks

f The equation of a least squares line of the form *weight* = $a + b \times$ *height* can be found from CAS. Write down the equation with a and b, correct to two decimal places.   2 marks

g Use the equation to predict the weight of the player who is 202 cm tall. Write your answer correct to two decimal places.   1 mark

h Calculate the residual for the prediction from part **g**.   2 marks

i Use CAS to complete the following table.   2 marks

Height (cm)	Weight (kg)	Predicted weight	Residual
181	85	85.3	−0.3
186	91	88.2	2.8
202	95		
190	92		
195	95		
189	94		
182	93		
192	86		
178	75		

j Draw the residual plot.   1 mark

k What does the residual plot tell us about the suitability of the linear model?   1 mark

113 VCAA 2009 Exam 2 (15 marks)

Question 113.1 (3 marks) ©VCAA 2009 2CQ1

Table 1 shows the number of rainy days recorded in a high rainfall area for each month during 2008.

Table 1

Month	Number of rainy days (mm)
January	12
February	8
March	12
April	14
May	18
June	18
July	20
August	19
September	17
October	16
November	15
December	13

The dot plot below displays the distribution of the number of rainy days for the 12 months of 2008.

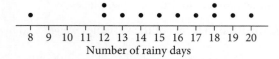

8 9 10 11 12 13 14 15 16 17 18 19 20
Number of rainy days

a **Circle** the dot on the dot plot that represents the number of rainy days in April 2008.   1 mark

b For the year 2008, determine

   i the median number of rainy days per month   1 mark

   ii the percentage of months that have more than 10 rainy days. Write your answer
     correct to the nearest per cent.   1 mark

Question 113.2 (4 marks) ©VCAA 2009 2CQ2

The time series plot below shows the rainfall (in mm) for each month during 2008.

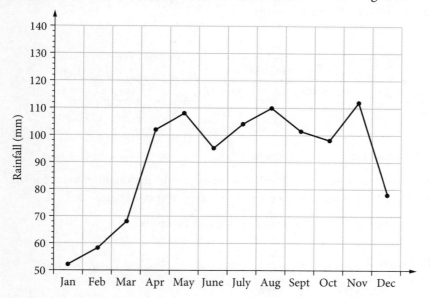

a Which month had the highest rainfall?   1 mark

b Use three-median smoothing to smooth the time series. Plot the smoothed time series
on the plot above. Mark each smoothed data point with a cross (**x**).   2 marks

c Describe the general pattern in rainfall that is revealed by the smoothed time series plot. 1 mark

Question 113.3 (5 marks) ©VCAA 2009 2CQ3

The scatterplot below shows the *rainfall* (in mm) and the *percentage of clear days* for each
month of 2008.

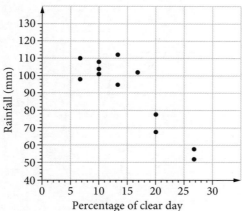

An equation of the least squares regression line for this data set is

$$rainfall = 131 - 2.68 \times percentage\ of\ clear\ days$$

a Draw this line on the scatterplot.   1 mark

b Use the equation of the least squares regression line to predict the rainfall for a month
with 35% of clear days. Write your answer in mm, correct to one decimal place.   1 mark

c The coefficient of determination for this data set is 0.8081.

   i Interpret the coefficient of determination in terms of the variables *rainfall* and
percentage of clear days.   1 marks

   ii Determine the value of Pearson's correlation coefficient. Write your answer correct
to three decimal places.   2 marks

Question 113.4 (3 marks) ©VCAA 2009 2CQ4

a Table 2 shows the seasonal indices for rainfall in summer, autumn and winter. Complete the table by calculating the seasonal index for spring.   1 mark

Table 2

Seasonal indices			
summer	**autumn**	**winter**	**spring**
0.78	1.05	1.07	

b In 2008, a total of 188 mm of rain fell during summer.

Using the appropriate seasonal index in Table 2, determine the de-seasonalised value for the summer rainfall in 2008. Write your answer correct to the nearest millimetre.   1 mark

c What does a seasonal index of 1.05 tell us about the rainfall in autumn?   1 mark

114 VCAA 2010 Exam 2 (15 marks)

Question 114.1 (5 marks) ©VCAA 2010 2CQ1 ●● ●

Table 1 shows the percentage of women ministers in the parliaments of 22 countries in 2008.

Table 1

Country	Percentage of women ministers
Norway	56
Sweden	48
France	47
Spain	44
Switzerland	43
Austria	38
Denmark	37
Iceland	36
Germany	33
Netherlands	33
New Zealand	32
Australia	24
Italy	24
United States	24
Belgium	23
United Kingdom	23
Ireland	21
Liechtenstein	20
Canada	16
Luxembourg	14
Japan	12
Singapore	0

a What proportion of these 22 countries have a higher percentage of women ministers in their parliament than Australia?   1 mark

b Determine the median, range and interquartile range of this data.   2 marks

The ordered stem plot below displays the distribution of the percentage of women ministers in parliament for 21 of these countries. The value for **Canada** is missing.

Stem 10(s)	Leaf (units)
0	0
1	2 4
2	0 1 3 3 4 4 4
3	2 3 3 6 7 8
4	3 4 7 8
5	6

c Complete the stem plot above by adding the value for Canada.   1 mark

d Both the median and the mean are appropriate measures of centre for this distribution. Explain why.   1 mark

Question 114.2 (4 marks) ©VCAA 2010 2CQ2 ●●●

In the scatterplot below, average annual *female income*, in dollars, is plotted against average annual *male income*, in dollars, for 16 countries. A least squares regression line is fitted to the data.

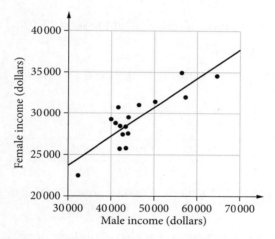

The equation of the least squares regression line for predicting female income from male income is

$$female\ income = 13\,000 + 0.35 \times male\ income$$

a What is the independent (explanatory) variable?   1 mark

b Complete the following statement by filling in the missing information.

From the least squares regression line equation, it can be concluded that, for these countries, on average, female income increases by $\boxed{}$ for each $1000 increase in male income.   1 mark

c **i** Use the least squares regression line equation to predict the average annual female income (in dollars) in a country where the average annual male income is $15 000.   1 mark

ii The prediction made in part **c i** is not likely to be reliable. Explain why.   1 mark

Question 114.3 (6 marks)   ©VCAA 2010 2CQ3   ●●●

Table 2 shows the Australian gross domestic product (GDP) per person, in dollars, at five yearly intervals for the period 1980 to 2005.

Table 2

Year	1980	1985	1990	1995	2000	2005
GDP	20 900	22 300	25 000	26 400	30 900	33 800

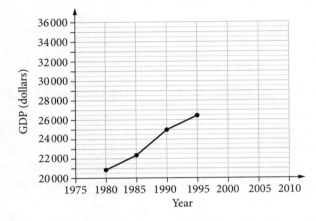

a Complete the time series plot above by plotting the GDP for the years 2000 and 2005.   1 mark

b Briefly describe the general trend in the data.   1 mark

In Table 3, the variable *year* has been rescaled using 1980 = 0, 1985 = 5 and so on. The new variable is *time*.

Table 3

Year	1980	1985	1990	1995	2000	2005
Time	0	5	10	15	20	25
GDP	20 900	22 300	25 000	26 400	30 900	33 800

c Use the variables *time* and *GDP* to write down the equation of the least squares regression line that can be used to predict *GDP* from *time*. Take *time* as the independent variable.   2 marks

d In the year 2007, the *GDP* was $34 900. Find the error in the prediction if the least squares regression line calculated in part **c** is used to predict *GDP* in 2007.   2 marks

115 VCAA 2012 Exam 2 (15 marks)

Question 115.1 (3 marks) ©VCAA 2012 2CQ1 ●●

The dot plot below displays the maximum daily temperature (in °C) recorded at a weather station on each of the 30 days in November 2011.

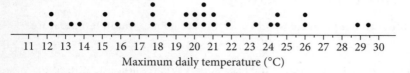

Maximum daily temperature (°C)

a   From this dot plot, determine

    i the median maximum daily temperature, correct to the nearest degree   1 mark

    ii the percentage of days on which the maximum temperature was less than 16°C.
        Write your answer, correct to one decimal place.   1 mark

Records show that the **minimum** daily temperature for November at this weather station is approximately normally distributed with a mean of 9.5°C and a standard deviation of 2.25°C.

b Determine the percentage of days in November that are expected to have a minimum daily temperature less than 14°C at this weather station.
Write your answer, correct to one decimal place.   1 mark

Question 115.2 (7 marks) ©VCAA 2012 2CQ2 ●●●

The maximum temperature and the minimum temperature at this weather station on each of the 30 days in November 2011 are displayed in the scatterplot below.

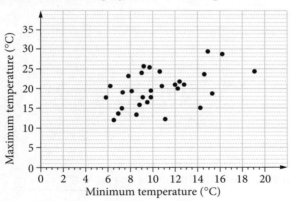

The correlation coefficient for this data set is $r = 0.630$.

The equation of the least squares regression line for this data set is

$$\textit{maximum temperature} = 13 + 0.67 \times \textit{minimum temperature}$$

a   **Draw** this least squares regression line on the scatterplot above.   1 mark

b   Interpret the vertical intercept of the least squares regression line in terms of maximum temperature and minimum temperature.   1 mark

c   Describe the relationship between the maximum temperature and the minimum temperature in terms of strength and direction.   1 mark

d Interpret the slope of the least squares regression line in terms of maximum temperature and minimum temperature.   1 mark

e Determine the percentage of variation in the maximum temperature that may be explained by the variation in the minimum temperature.

Write your answer, correct to the nearest percentage.   1 mark

On the day that the minimum temperature was 11.1°C, the actual maximum temperature was 12.2°C.

f Determine the residual value for this day if the least squares regression line is used to predict the maximum temperature.

Write your answer, correct to the nearest degree.   2 marks

Question 115.3 (2 marks)   ©VCAA 2012 2CQ3

A weather station records the wind speed and the wind direction each day at 9.00 am.

The wind speed is recorded, correct to the nearest whole number.

The parallel boxplots below have been constructed from data that was collected on the 214 days from June to December in 2011.

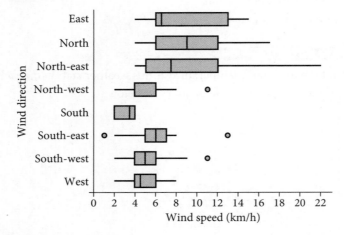

a Complete the following statements.

The wind direction with the lowest recorded wind speed was [   ].

The wind direction with the largest range of recorded wind speeds was [   ].   1 mark

b The wind blew from the south on eight days.

Reading from the parallel boxplots above we know that, for these eight wind speeds, the

- first quartile, $Q_1 = 2$ km/h

- median, $M = 3.5$ km/h

- third quartile, $Q_3 = 4$ km/h

Given that the eight wind speeds were recorded to the nearest whole number, write down the eight wind speeds.   1 mark

Question 115.4 (3 marks) ©VCAA 2012 2CQ4

The wind speeds (in km/h) that were recorded at the weather station at 9.00 am and 3.00 pm respectively on 18 days in November are given in the table below. A scatterplot has been constructed from this data set.

Wind speed (km/h)	
9.00 am	**3.00 pm**
2	2
4	6
4	7
4	4
13	11
6	7
3	3
16	10
6	7
13	8
11	9
2	4
7	8
5	5
8	6
6	7
19	11
9	9

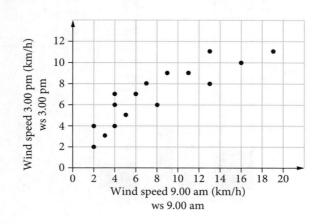

Let the wind speed at 9.00 am be represented by the variable *ws9.00am* and the wind speed at 3.00 pm be represented by the variable *ws3.00pm*.

The relationship between *ws9.00am* and *ws3.00pm* shown in the scatterplot above is non-linear.

A **squared transformation** can be applied to the variable *ws3.00pm* to linearise the data in the scatterplot.

a Apply the squared transformation to the variable *ws3.00pm* and determine the equation of the least squares regression line that allows $(ws3.00pm)^2$ to be predicted from *ws9.00am*. In the boxes provided, write the coefficients for this equation, correct to one decimal place.   2 marks

$$(ws3.00pm)^2 = \boxed{} + \boxed{} \times ws9.00am$$

b Use this equation to predict the wind speed at 3.00 pm on a day when the wind speed at 9.00 am is 24 km/h. Write your answer, correct to the nearest whole number.   1 mark

116 VCAA 2014 Exam 2 (15 marks)

Question 116.1 (3 marks) ©VCAA 2014 2CQ1 ●●

The segmented bar chart below shows the age distribution of people in three countries, Australia, India and Japan, for the year 2010.

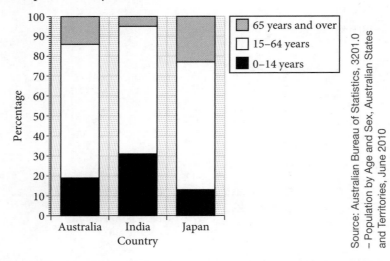

a Write down the percentage of people in Australia who were aged 0–14 years in 2010. Write your answer, correct to the nearest percentage.   1 mark

b In 2010, the population of Japan was 128 000 000.

   How many people in Japan were aged 65 years and over in 2010?   1 mark

c From the graph above, it appears that there is no association between the percentage of people in the 15–64 age group and the country in which they live.

   Explain why, quoting appropriate percentages to support your explanation.   1 mark

Question 116.2 (6 marks) ©VCAA 2014 2CQ2 ●●●

The scatterplot below shows the *population* and *area* (in square kilometres) of a sample of inner suburbs of a large city.

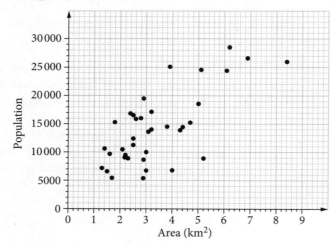

The equation of the least squares regression line for the data in the scatterplot is

$$population = 5330 + 2680 \times area$$

a Write down the dependent (response) variable.   1 mark

b Draw the least squares regression line on the **scatterplot**.   1 mark

c Interpret the slope of this least squares regression line in terms of the variables *area* and *population*.

2 marks

d Wiston is an inner suburb. It has an area of $4\,km^2$ and a population of 6690. The correlation coefficient, *r*, is equal to 0.668

   i Calculate the residual when the least squares regression line is used to predict the population of Wiston from its area.

1 mark

   ii What percentage of the variation in the population of the suburbs is explained by the variation in area? Write your answer, correct to one decimal place.

1 mark

Question 116.3 (2 marks) ©VCAA 2014 2CQ3 ●●●

The scatterplot and table below show the *population*, in thousands, and the *area*, in square kilometres, for a sample of 21 outer suburbs of the same city.

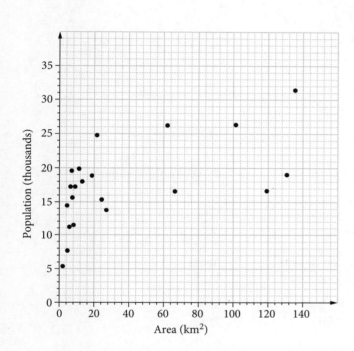

Area (km^2)	Population (thousands)
1.6	5.2
4.4	14.3
4.6	7.5
5.6	11.0
6.3	17.1
7.0	19.4
7.3	15.5
8.0	11.3
8.8	17.1
11.1	19.7
13.0	17.9
18.5	18.7
21.3	24.6
24.2	15.2
27.0	13.6
62.1	26.1
66.5	16.4
101.4	26.2
119.2	16.5
130.7	18.9
135.4	31.3

In the outer suburbs, the relationship between *population* and *area* is non-linear.

A **log** transformation can be applied to the variable *area* to linearise the scatterplot.

a Apply the **log** transformation to the data and determine the equation of the least squares regression line that allows the population of an outer suburb to be predicted from the logarithm of its area.

Write the slope and intercept of this regression line in the boxes provided below. Wrie your answers, correct to one decimal place.

1 mark

$$population = \boxed{} + \boxed{} \times \log_{10}(area)$$

b Use this regression equation to predict the population of an outer suburb with an area of 90 km².

Write your answer, correct to the nearest one thousand people.   1 mark

Question 116.4 (4 marks) ©VCAA 2014 2CQ4 ●●●

The scatterplot below shows the *population density*, in people per square kilometre, and the *area*, in square kilometres, of 38 inner suburbs of the same city.

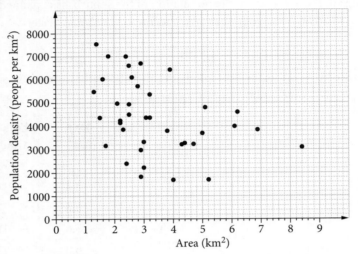

For this scatterplot, $r^2 = 0.141$.

a Describe the association between variables *population density* and *area* for these suburbs in terms of strength, direction and form.   1 mark

b The mean and standard deviation of the variables *population density* and *area* for these 38 inner suburbs are shown in the table below.

	Population density (people per km²)	**Area (km²)**
Mean	4370	3.4
Standard deviation	1560	1.6

   i One of these suburbs has a population density of 3082 people per square kilometre. Determine the standard z-score of this suburb's population density.

   Write your answer, correct to one decimal place.   1 mark

Assume the areas of these inner suburbs are approximately normally distributed.

   ii How many of these 38 suburbs are **expected** to have an area that is two standard deviations or more above the mean?

   Write your answer, correct to the nearest whole number.   1 mark

   iii How many of these 38 inner suburbs **actually** have an area that is two standard deviations or more above the mean?   1 mark

117 VCAA 2015 Exam 2 (15 marks)

Question 117.1 (3 marks) ©VCAA 2015 2CQ1 ●●

The histogram below shows the distribution of life expectancy of people for 183 countries.

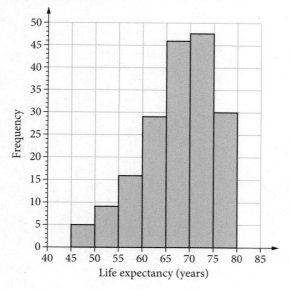

a For this distribution, the modal interval is [   ].   1 mark

b In how many of these countries is life expectancy less than 55 years?   1 mark

c In what percentage of these 183 countries is life expectancy between 75 and 80 years?

   Write your answer correct to one decimal place.   1 mark

Question 117.2 (3 marks) ©VCAA 2015 2CQ2 ●●

The parallel boxplots below compare the distribution of life expectancy for 183 countries for the years 1953, 1973 and 1993.

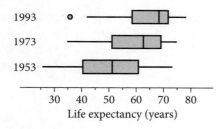

a Describe the shape of the distribution of life expectancy for 1973.   1 mark

b Explain why life expectancy for these countries is associated with the year. Refer to specific statistical values in your answer.   2 marks

Question 117.3 (3 marks) ©VCAA 2015 2CQ3 ●●

The scatterplot below plots male life expectancy (*male*) against female life expectancy (*female*) in 1950 for a number of countries. A least squares regression line has been fitted to the scatterplot as shown.

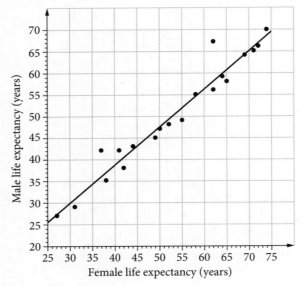

The slope of this least squares regression line is 0.88.

a Interpret the slope in terms of the variables *male* life expectancy and *female* life expectancy.   1 mark

The equation of this least squares regression line is

$$male = 3.6 + 0.88 \times female$$

b In a particular country in 1950, *female* life expectancy was 35 years.

Use the equation to predict *male* life expectancy for that country.   1 mark

c The coefficient of determination is 0.95.

Interpret the coefficient of determination in terms of male life expectancy and female life expectancy.   1 mark

Question 117.4 (2 marks)   ©VCAA 2015 2CQ4   ●●●

The table below shows male life expectancy (*male*) and female life expectancy (*female*) for a number of countries in 2013. The scatterplot has been constructed from this data.

Life expectancy (in years) in 2013	
Male	**Female**
80	85
60	62
73	80
70	71
70	78
78	83
77	80
65	69
74	77
70	78
75	81
58	60
80	86
69	73
79	84
72	81
78	85
72	79
77	81
71	80

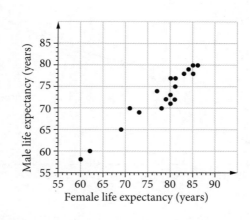

a Use the scatterplot to describe the association between *male* life expectancy and *female* life expectancy in terms of strength, direction and form.   1 mark

b Determine the equation of a least squares regression line that can be used to predict *male* life expectancy from *female* life expectancy for the year 2013.

Complete the equation for the least squares regression line below by writing the intercept and slope in the boxes provided.

Write these values correct to two decimal places.   1 mark

$$male = \boxed{} + \boxed{} \times female$$

Question 117.5 (4 marks) ©VCAA 2015 2CQ5 ●●○

The time series plot below displays the *life expectancy*, in years, of people living in Australia and the United Kingdom (UK) for each *year* from 1920 to 2010.

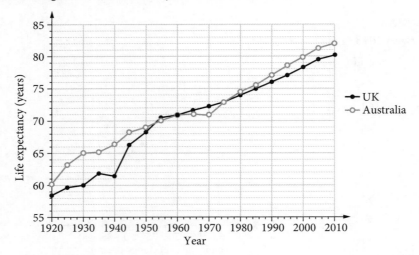

a By how much did *life expectancy* in Australia increase during the period 1920 to 2010? Write your answer correct to the nearest year.   1 mark

b In 1975, the life expectancies in Australia and the UK were very similar.

From 1975, the gap between the life expectancies in the two countries increased, with people in Australia having a longer life expectancy than people in the UK.

To investigate the difference in life expectancies, least squares regression lines were fitted to the data for both Australia and the UK for the period 1975 to 2010.

The results are shown below.

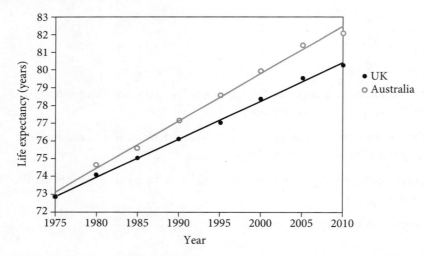

The equations of the least squares regression lines are as follows.

Australia: *life expectancy* = −451.7 + 0.2657 × *year*

UK: *life expectancy* = −350.4 + 0.2143 × *year*

i Use these equations to predict the difference between the life expectancies of Australia and the UK in 2030.

Give you answer correct to the nearest year.   2 marks

ii Explain why this prediction may be of limited reliability.   1 mark

Chapter 2 Recursion and financial modelling
Area of Study 2: Discrete mathematics

Content summary notes

Recursion and financial modelling

Depreciation of assets

- Percentage increase and decrease
- Depreciation of the value of an asset

Simple interest and compound interest investments and loans

- Appreciation of value (increase in value) with simple interest
- Appreciation of value (increase in value) with compound interest
- Compound interest loans
- Effective rate of interest

Reducing balance loans

- Using of the Finance solver on CAS
- Comparison of loans
- Interest only loans

Annuities and perpetuities

- Annuities
- Perpetuities

Compound interest in investment with periodic and equal additions to the principal

Area of Study 2 Outcome 1

On completing this outcome, you should be able to:

- the facts, concepts and techniques associated with data analysis, recursion and financial modelling, matrices and networks and decision mathematics

- standard models studied in data analysis, recursion and financial modelling, matrices, and networks and decision mathematics, and their areas of application

- general formulation of the concepts, techniques and models studied in data analysis, recursion and financial modelling, matrices, and networks and decision mathematics

- assumptions and conditions underlying the use of the concepts, techniques and models associated with data analysis, recursion and financial modelling, matrices, and networks and decision mathematics

Although you should become familiar with all of the key skills not all the skills are required for the exam.

- identify, recall and select facts, concepts, models and techniques needed to investigate and analyse statistical features of a data set with several variables that can include time series data

- select and implement standard financial models to investigate and analyse a financial or mathematically equivalent non-financial situation that requires the use of increasingly sophisticated models to complete the analysis

- identify, recall and select the mathematical concepts, models and techniques needed to solve an extended problem or conduct an investigation in a variety of contexts related to matrices and networks and decision mathematics

- interpret and report the results of a statistical investigation or of completing a modelling or problem-solving task in terms of the context under consideration, including discussing the assumptions in application of these models

VCE Mathematics Study Design 2023–2027 p. 91, © VCAA 2022

2.1 Depreciation of assets

A **first order linear recurrence relation** is a relation that links each **term** of a sequence to the previous term. Usually, the first term of the sequence is stated.

If u_n is the nth term of a sequence, then u_{n+1} is the term after u_n.

General form, where b and c are constants:

$u_{n+1} = bu_n + c$

The value of $u_0 = a$ is usually given, where a is a constant.

Subsequent terms can be generated from the recurrence relation:

$u_1 = bu_0 + c$

$u_2 = bu_1 + c$

$u_3 = bu_2 + c$,   etc.

First order recurrence relations generate sequences that can exhibit linear growth (arithmetic sequences), geometric growth (geometric sequences) or neither.

Arithmetic sequences are sequences where each new term after the first term is formed by adding a constant number to the previous term. The recurrence relation has the form:

$u_0 = a, \; u_{n+1} = bu_n + c$

where a and c are constants and $b = 1$.

Terms of an arithmetic sequence can represent the accumulation where **simple interest** is applied.

The **graph of an arithmetic sequence**, where the constant $b = 1$, is a set of discrete points that form a straight line.

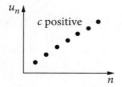
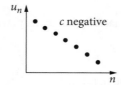

Geometric sequences are sequences where each new term after the first term is formed by multiplying the previous term by a constant. The recurrence relation has the form:

$u_0 = a, \; u_{n+1} = bu_n + c$

where a and b are constants, $b \neq 1$ and $c = 0$.

Terms of a geometric sequence can represent the accumulation where **compound interest** is applied.

The **graph of a geometric sequence** is a set of discrete points that do not form a straight line. Two basic shapes are as illustrated for different values of b.

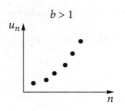
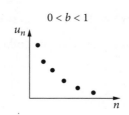

2.1.1 Percentage increase and decrease

2.1.1.1 Percentage increase

If a **percentage increase** of n% takes place, then the original value is multiplied by $\left(1 + \dfrac{n}{100}\right)$, which is known as the multiplying factor.

For example, to account for percentage increases of 50%, 10% and 5%, the factors to multiply by are 1.5, 1.1 and 1.05 respectively.

Example 1

An asset's value is credited monthly and is increasing by 6.6% p.a. What is the monthly multiplying factor?

Solution

6.6% p.a. is equal to $\dfrac{6.6}{12}$ or 0.55% per month.

monthly multiplying factor $= 1 + \dfrac{0.55}{100} = 1.0055$

2.1.1.2 Percentage decrease

The multiplying factor is $\left(1 - \dfrac{n}{100}\right)$, where n is the percentage decrease.

For example, to account for **percentage decreases** of 50%, 10% and 5%, the factors to multiply by are 0.5, 0.9 and 0.95 respectively.

Example 2

An asset's value is calculated quarterly but is decreasing by 14% p.a. What is the quarterly multiplying factor?

Solution

14% p.a. is equal to $\dfrac{14}{4}$ or 3.5% per quarter.

quarterly multiplying factor $= 1 - \dfrac{3.5}{100} = 0.965$

2.1.2 Depreciation of the value of an asset

Depreciation is the decrease in **value of an asset** over time.

Commonly used terms:

book value: the current value of an item that is being depreciated.

useful life: the time during which the item produces revenue for a business.

scrap value: the value remaining once the period of useful life has expired.

2.1.2.1 Flat rate depreciation

Also known as **straight-line depreciation**, this is **flat rate depreciation** is calculated each year as a percentage, r, of the original value, P. This means that the item decreases in value by the same amount each year. The annual depreciation, d, is given by:

$d = \dfrac{rP}{100}$ each year

As a **recurrence relation**, the value of the asset **after the nth year** is given by:

$u_0 = P, u_{n+1} = u_n - d$

A **graph** of the book value versus time is a series of points in a straight line, with a negative slope.

A **formula** to calculate the book value **after n years**:

book value $= P - nd$

Example 3

An equipment, bought for \$11 500 decreases by 12% of the original value each year. After how many years is its value less than \$5000?

Solution

depreciation $=$ by 12% of \$11 500

$\qquad = \dfrac{12}{100} \times \$11\,500$

$\qquad = \$1380$ each year

So we need to solve: $11\,500 - 1380 \times n < 5000$

$\qquad\qquad\qquad\qquad 6500 < 1380n$

$\qquad\qquad\qquad\qquad\quad n > 4.71$ years

After 5 years, the value is first less than \$5000. (This answer can easily be found by iteration on CAS.)

2.1.2.2 Reducing balance depreciation

Reducing balance depreciation is depreciation that is calculated each year as a percentage, r, of the book value. This means that the item decreases in value by a smaller amount each year.

Therefore, the values of the item each year form the terms of a geometric sequence, where P is the original value of the item and $R = 1 - \dfrac{r}{100}$ is the depreciation multiplying factor each year.

As a **recurrence relation**, the value **after the nth year** is given by

$u_0 = P, u_{n+1} = Ru_n$

A **graph** of the book value versus time is a decreasing curve.

A **formula** to calculate the book value **after n years**:

book value $= PR^n$

Example 4

A recurrence relation for the depreciation of office furniture is given as $u_0 = 16\,000$, $u_{n+1} = 0.82u_n$ where u_n is the value of the furniture after n years.

a State the initial cost of the furniture.

b State the percentage decrease each year.

c Showing recursive calculations, find the value of the furniture after 4 years.

Solution

a The initial cost (u_0) is $\$16\,000$.

b $0.82 = 1 - \dfrac{r}{100}$, where r = percentage decrease

$\dfrac{r}{100} = 1 - 0.82$

$\qquad r = 0.18 \times 100$

$\qquad\quad = 18\%$

c $u_1 = 0.82 \times u_0 = 0.82 \times 16\,000 = 13\,120$

$u_2 = 0.82 \times u_1 = 0.82 \times 13\,120 = 10\,758.40$

$u_3 = 0.82 \times u_2 = 0.82 \times 10\,758.4 = 8821.89$

$u_4 = 0.82 \times u_3 = 0.82 \times 8821.89 = 7233.95$

The value of the furniture after 4 years is $\$7233.95$.

2.1.2.3 Unit cost depreciation

Unit cost depreciation is where the item depreciates through the amount it is used rather than its age.

A cost, x, per unit of use is given.

As a recurrence relation, the value of the asset after the nth use is given by

$\qquad u_0 = P,\ u_{n+1} = u_n - nx$

If P is the original cost of the item, then a formula for the value of the asset after n units of use is given by:

$\qquad$ book value $= P - nx$

Example 5

A photocopier bought for $\$20\,000$ depreciates by 1 cent for each copy made. In the first year, $400\,000$ copies are made. What is its value after one year?

Solution

book value after 1 year $= 20\,000 - 400\,000 \times 0.01$

$\qquad\qquad\qquad\qquad\quad = \$16\,000$

2.2 Simple interest and compound interest investments and loans

2.2.1 Appreciation of value (increase in value) with simple interest

Simple interest is also known as **flat rate interest**. Simple interest is calculated as a percentage of the initial value of an asset and is the same amount each time period.

A formula to calculate the interest after T years is:

$$I = \frac{PrT}{100}$$

where

I is the interest in dollars
P is the principal in dollars
r is the percentage interest rate per annum
T is the term in years.

Example 6

An asset valued at $12 000 is being credited with simple interest of 8.4% p.a. How much interest is earned each year? Calculate this value using a recurrence relation, and then verify this result using the simple interest formula.

Solution

8.4% of $12 000 = $1008 in interest each year

A recurrence relation to calculate the value of this investment after the nth year:

$u_0 = 12\,000,\ u_{n+1} = u_n + 1008$

From this we can calculate

$u_1 = u_0 + 1008 = 12\,000 + 1008 = 13\,008$

$u_2 = u_1 + 1008 = 13\,008 + 1008 = 14\,016$, etc.

You can verify this result using the simple interest formula.

After two years, the interest earned is:

$$I = \frac{12\,000 \times 8.4 \times 2}{100} = \$2016$$

Value of the investment after two years = $12 000 + $2016 = $14 016

CHAPTER 2

2.2.2 Appreciation of value (increase in value) with compound interest

With compound interest **appreciation** of an investment, the interest is added to the original value after each time period. This effectively creates a new value upon which the next interest calculation is based.

The values of the investment after each time period form the terms of a geometric sequence.

A formula to calculate the **amount accumulated**, A, after n compounding periods is:

$A = PR^n$

where $R = 1 + \dfrac{r}{100}$

P is the principal
R is the multiplying factor
r is the percentage interest rate per time period
n is the number of compounding periods.

Example 7

An investment of \$10 000 is being credited with compound interest at 6.6% p.a., calculated yearly. Find the value of the investment after each of the first three years. Calculate this value using a recurrence relation, and then verify this result using the compound interest formula.

Solution

6.6% p.a. means a multiplying factor of 1.066.

After the first year, the value of the investment is

$\$10\,000 \times 1.066 = \$10\,660$

After the second year, the value of the investment is

$\$10\,660 \times 1.066 = \$11\,363.56$

After the third year, the value of the investment is

$\$11\,363.56 \times 1.066 = \$12\,113.55$

A **first order recurrence relation** to calculate the value, Vn, of this investment **after the nth year**:

$V_0 = 10\,000, \quad V_{n+1} = 1.066V_n$

From this, we can calculate

$V_1 = 1.066V_0 = 1.066 \times 10\,000 = 10\,660$

$V_2 = 1.066V_1 = 1.066 \times 10\,660 = 11\,363.56$

$V_3 = 1.066V_2 = 1.066 \times 11\,363.56 = 12\,113.55$

You can verify this result using the formula $A = PR^n$.

The value of the investment after 3 years is:

$A = 10\,000 \times 1.066^3 = \$12\,113.55$

interest earned $= A - P = \$12\,113.55 - \$10\,000 = \$2113.55$

Example 8

Klaus invested $15 000 at 6.4% per annum, compounding quarterly. Determine the interest earned if the investment was made for 4 years.

Solution

Interest rate is 6.4% per annum, therefore $r = \dfrac{6.4}{4} = 1.6\%$ per quarter.

Number of compounding periods, $n = 4 \times 4 = 16$.

$A = PR^n$, where $R = 1 + \dfrac{r}{100}$

$A = 15\,000 \times \left(1 + \dfrac{1.6}{100}\right)^{16} = \$19\,337.07$

Interest earned $= A - P = \$19\,337.07 - \$15\,000 = \$4337.07$

The graph below shows the value of the investment over the four years.

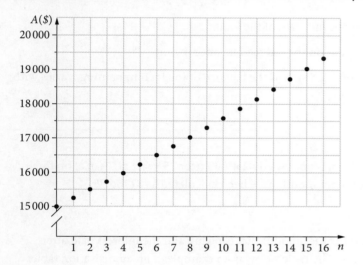

Note
The points are not in a straight line.

Note
In the long term, an investment, at the same rate, will earn less with simple interest than with compound interest.

2.2.3 Compound interest loans

Example 9

Janet borrowed $5000 from the bank. She is to repay the loan, plus interest, in five years' time. Interest is being charged at 5% p.a. compounding yearly.

a Calculate the amount that Janet will be repaying.

b Will Janet be paying more or less if the interest is calculated quarterly?

Solution

a Use the formula $A = PR^n$, where $R = 1 + \dfrac{r}{100} = 1 + \dfrac{5}{100}$

Amount that Janet will owe $= 5000 \times 1.05^5 = \$6381.41$

b If the interest is calculated quarterly, then $R = 1 + \dfrac{\frac{5}{4}}{100}$ (1.0125) and there will be 20 compounding periods.

Amount that Janet will owe $= 5000 \times 1.0125^{20} = \$6410.19$

She will owe the bank $6410.19 if interest is calculated quarterly. Hence Janet will be paying more.

2.2.4 Effective rate of interest

An **effective rate of interest** differs from the nominal rate of interest and is used to compare investment returns and the cost of loans when interest is paid or charged, for example, daily, monthly, quarterly.

$$\text{effective interest rate per annum} = r_{effective} = \left[\left(1 + \frac{r}{100n}\right)^n - 1\right] \times 100\%$$

where r is the nominal interest rate and n is the number of compounding periods per annum.

2.3 Reducing balance loans

Characteristics of a **reducing balance loan**:

- The term, rate of interest, amount of the loan and repayment amount are set at the start of the loan. Any of these variables can change throughout the term of the loan but they will affect all or some of the other variables.

- Regular repayments are made throughout the term of the loan.

- The interest is calculated each time period on the reduced balance owing.

- With each repayment the proportion of interest, and the balance of the loan remaining, varies with more interest being paid early in the loan.

Example 10

Tommy has just purchased a new home. He has borrowed $250 000 over 25 years at 7% p.a. on the reducing monthly balance. His repayments per month have been calculated as $1766.95.

Solution

The following step-by-step calculations show the amount that is owing and the interest paid after each of the first four compounding periods. The table below is an example of an amortisation table.

Payment number	Payment	Interest	Principal reduction	Balance
0	0.00	0.00	0.00	250 000.00
1	1766.95	1458.33	308.62	249 691.38
2	1766.95	1456.53	310.41	249 380.97
3	1766.95	1454.72	312.23	249 068.74
4	1766.95	1452.90	314.05	248 754.69

The interest is calculated as balance $\times \dfrac{\frac{7}{12}}{100}$.

Each new balance is calculated as:

   previous balance − principal reduction

It can be seen from the table that the balance and also the amount of interest paid is decreasing with each repayment.

A recurrence relation for this reducing balance loan:

   $A_0 = 250\,000$, $A_{n+1} = 1.005\,833\,33 A_n - 1766.95$

where A_n is the amount owing after n years.

> **Note**
>
> $1 + \dfrac{\frac{7}{12}}{100} = 1.005\,833\,33\ldots$

2.3.1 Using the Finance solver on CAS

In examinations, answers can be determined directly from CAS.

Using the previous example, the amount that Tommy needs to repay each month can be determined using the **Finance solver** in CAS. To find this amount, we assume that the amount owing (FV) will be zero after 300 months (25 years).

N is the number of time periods (*n* in the formula) = 25 × 12 = 300

I% – percentage interest rate per annum = 7%

PV – the present value of the loan = 250 000

PMT – the repayment made each time period (unknown)

FV – the future value of the loan = 0

P/Y – the number of payments per year

C/Y – number of compounding periods per year
(adjusts when P/Y is selected) = 12 (monthly)

Be careful with the use of signs (positive or negative) for PV, PMT and FV.

- PV (present value) is the amount borrowed. Since it is money coming in, it is considered as a positive value.

- PMT (payments) are outgoing amounts so are considered as negative values.

- FV (future value) is considered as a negative value (if not 0) as this would be outgoing if the loan was being fully repaid.

Note: this screenshot is from the Amortisation tab rather than compound interest tab. PM1 and PM2 indicate the length of time from start to finish (i.e. from month 1 to month 300).

The sum of the interest over 300 months is $280 084.40. A negative figure denotes this is an outgoing amount.

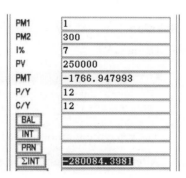

total amount paid for this loan = 300 × $1766.95 = $530 085

Total interest paid can be calculated as follows:

number of payments × value of payment – amount borrowed = 300 × $1766.95 – 250 000
$$= \$280\,085.00$$

2.3.2 Comparison of loans

Changes to one of the variables will affect the length of the term and the total interest paid.

- An increase in the interest rate will increase the term of the loan and increase the total interest to be paid OR it will increase the amount of the repayment.

- An increase in the frequency of payments will decrease the term of the loan and decrease the total interest to be paid OR decrease the amount of the repayment.

- An increase in the amount repaid per period will decrease the term of the loan and decrease the total interest to be paid.

2.3.3 Interest only loans

An **interest-only loan** is often obtained for investment in an asset where a return relies on the increase in value of the asset over time. The repayments for an interest-only loan will be lower than those for a reducing balance loan as the balance is not reducing. The repayment amount will be for the interest owing over the time period.

Example 11

Marion has an interest-only loan of $350 000 for an investment property, over ten years. The interest being charged is 6.4% p.a. and she will make her repayments monthly.

a What is the repayment amount per month?

b How much will Marion owe after ten years?

Solution

a Using CAS:

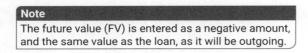

> **Note**
> The future value (FV) is entered as a negative amount, and the same value as the loan, as it will be outgoing.

The repayment amount is $1866.67 per month.

b After ten years, Marion will still owe $350 000.

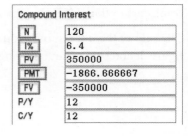

Compound Interest	
N	120
I%	6.4
PV	350000
PMT	−1866.666667
FV	−350000
P/Y	12
C/Y	12

2.4 Annuities and perpetuities

2.4.1 Annuities

An **annuity** is a **compound interest investment** that has periodic payments made from the investment. Annuities are often used by people in their retirement who have a lump sum to invest and who want to receive a regular income of a set amount.

Example 12

Helen is going to invest \$80 000 in an annuity that credits the investment with 5% p.a. interest compounding yearly, for 6 years. The annuity is designed to provide \$15 000 per year for Helen.

Show step-by-step calculations for the 6 years of the investment. How much is remaining in the account at the end of 6 years?

Solution

Year	Payment	Interest	Principal reduction	Balance
0	0.00	0.00	0.00	80 000.00
1	15 000.00	4000.00	11 000.00	69 000.00
2	15 000.00	3450.00	15 000.00	57 450.00
3	15 000.00	2872.50	15 000.00	45 322.50
4	15 000.00	2266.13	15 000.00	32 588.63
5	15 000.00	1629.43	15 000.00	19 218.06
6	15 000.00	960.90	15 000.00	5 178.96

There is \$5178.96 remaining at the end of 6 years.

This situation can be modelled by a **recurrence relation**, where V_n is the balance of the account after the nth year.

$$V_0 = 80\,000, \quad V_{n+1} = 1.05V_n - 15\,000$$

giving the values

$$V_1 = 1.05V_0 - 15\,000 = 69\,000$$

$$V_2 = 1.05V_1 - 15\,000 = 1.05 \times 69\,000 - 15\,000 = 57\,450$$

$$V_3 = 1.05V_2 - 15\,000 = 1.05 \times 57\,450 - 15\,000 = 45\,322.50$$

$$V_4 = 1.05V_3 - 15\,000 = 1.05 \times 45\,322.50 - 15\,000 = 32\,588.63$$

$$V_5 = 1.05V_4 - 15\,000 = 1.05 \times 32\,588.63 - 15\,000 = 19\,218.06$$

$$V_6 = 1.05V_5 - 15\,000 = 1.05 \times 19\,218.06 - 15\,000 = 5178.96$$

Example 13

Irina has \$480 000 to invest in an annuity that will credit her with compound interest of 4.8% on the reducing balance. Interest compounds monthly.

a How much will Irina receive each month from this investment if it is to last 30 years?

b How long will the investment last if she receives \$4000 per month?

Solution

a Using CAS, Irina will receive \$2518.39 per month if the annuity is to last for 30 years.

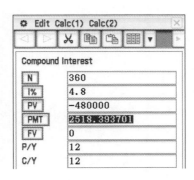

b If Irina receives $4000 per month, the investment will last for 163 months (13 years 7 months). The final payment will be $3221.80 if it is made at the end of the 163rd month.

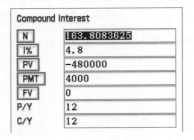

2.4.2 Perpetuities

A **perpetuity** is an investment that provides a regular payment (interest) on an indefinite basis without affecting the original value of the investment. The perpetuity is the interest earned by the principal in the specified time period.

The most likely use of a perpetuity is as a pension that a retired person will receive for the remainder of their life, leaving the original amount invested as part of an estate.

$$P = \frac{100Q}{r}$$

where

P = the initial principal invested
Q = perpetuity paid per time period
r = percentage interest rate per time period

Example 14

Albert invests $500 000 in a perpetuity at 4.9% per annum to provide income in his retirement. Determine the fortnightly amount he will receive.

Solution

$P = 500\,000,\ r = \dfrac{4.9}{26} = 0.188\,462$

$P = \dfrac{100Q}{r}$

Rearrange to make Q the subject:

$Q = \dfrac{Pr}{100} = \dfrac{500000 \times 0.188462}{100} = \$942.31$

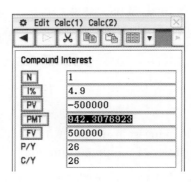

2.5 Compound interest investment with periodic and equal additions to the principal

This arises when money is invested under a compound interest arrangement and additional payments are made each time period into the account.

Example 15

$8000 is invested in an account that pays 4.8% p.a. interest, compounding monthly, and an additional $100 is deposited at the end of each month. How much is in the account after the 3rd month?

Solution

monthly interest rate $= \dfrac{4.8}{12} = 0.4\%$

monthly multiplying factor $= 1 + \dfrac{0.4}{100} = 1.004$

At the end of each month, the value of the investment will be (starting value = $8000)

1st month	$8000 \times 1.004 + 100 = 8132$
2nd month	$8132 \times 1.004 + 100 = 8264.53$
3rd month	$8264.53 \times 1.004 + 100 = 8397.59$

This can be modelled by a recurrence relation, where V_n is the amount in the account after the nth month.

$$V_0 = 8000, \quad V_{n+1} = 1.004V_n + 100$$

CAS can be used to find values of the investment after n compounding periods.

On CAS, the sign of the present value and the payment must both be the same. In this case, the original investment and the payments are considered to be 'outgoings' and so have negative signs.

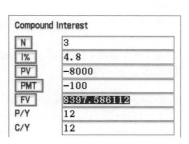

Compound Interest	
N	3
I%	4.8
PV	−8000
PMT	−100
FV	8397.586112
P/Y	12
C/Y	12

CHAPTER 2

Glossary

A+ DIGITAL
Revise this topic's key terms
and concepts by scanning
the QR code or typing the
URL into your browser.

https://get.ga/aplus-
vcegeneral-maths

amount accumulated An amount accumulated, A, after n compounding periods is given by $A = PR^n$, where $R = 1 + \dfrac{r}{100}$ and P is the principal, R is the multiplying factor, r is the percentage interest rate per time period and n is the number of compounding periods.

annuity A compound interest investment that has periodic payments made from the investment.

appreciation (of the value of an asset) The increase in value of an asset over time.

arithmetic sequences Sequences where each new term after the first is formed by adding a constant number to the previous term so the recurrence relation has the form $u_0 = a$, $u_{n+1} = bu_n + c$, where a and c are constant and $b = 1$.

book value The current value of an item that is being appreciated or depreciated.

Book value after n years for flat rate depreciation $= P - nd$

Book value after n years for reducing balance depreciation $= PR^n$

compound interest investment Periodic and equal additions to the principal when money is invested under a compound interest arrangement and additional payments are made each time period into the account.

depreciation (of the value of an asset) The decrease in value of an asset over time.

effective rate of interest A higher rate than the flat rate as it takes into account the reductions in the principal owed.

effective interest rate formula per annum

$$= r_{effective} = \left[\left(1 + \frac{r}{100n}\right)^n - 1\right] \times 100\%$$

where r is the nominal interest rate and n is the number of compounding periods per annum.

Finance solver (on CAS)

I% – percentage interest rate per annum (r in the formula)

PV – the present value of the loan (P in the formula)

PMT – the repayment made each time period (Q in the formula).

FV – the future value of the loan (A in the formula)

P/Y – the number of payments per year

C/Y – number of compounding periods per year

first order recurrence relations A relation that links each term of a sequence to the previous term. Usually, the first term of the sequence is stated. Generated sequences can exhibit linear growth (arithmetic sequences), geometric growth (geometric sequences) or neither.

flat rate depreciation (also called **straight-line depreciation**) is where depreciation is calculated each year as a percentage, r, of the original value, P. This means that the item decreases in value by the same amount $d = \dfrac{rP}{100}$ each year.

general form of depreciation of assets

$u_{n+1} = bu_n + c$, where $u_0 = a$ and b and c are constants

geometric sequence Sequences where each new term after the first is formed by multiplying the previous term by a constant number so the recurrence relation has the form $u_0 = a$, $u_{n+1} = bu_n$, where a and b are constants.

graph of an arithmetic sequence (where the constant $b = 1$) A set of discrete points that form a straight line.

graph of a geometric sequence A set of discrete points that do not form a straight line.

interest-only loan An asset where a return relies on the increase in value of the asset over time. The repayments for an interest-only loan will be lower than those for a reducing balance loan as the balance is not reducing. The repayment amount will be for the interest owing over the time period.

percentage decrease If a percentage decrease of $n\%$ takes place, then the original value is multiplied by $\left(1 - \dfrac{n}{100}\right)$ which is known as the multiplying factor.

percentage increase If a percentage increase of $n\%$ takes place, then the original value is multiplied by $\left(1 + \dfrac{n}{100}\right)$, which is known as the multiplying factor.

9780170465335

perpetuity An investment that provides a regular payment (interest) on an indefinite basis without affecting the original value of the investment. The interest earned by the principal in the specified time period.

reducing balance depreciation An amount calculated each year as a percentage, r, of the book value. This means that the item decreases in value by a smaller amount each year.

reducing balance loans

- The term, rate of interest, amount of the loan and repayment amount are set at the start of the loan.
- Regular repayments are made throughout the term of the loan.
- The interest is calculated each time period on the reduced balance owing.
- With each repayment the proportion of interest, and the balance of the loan remaining, varies with more interest being paid early in the loan.

scrap value The value remaining once the period of useful life of an asset has expired.

simple interest Also known as **flat rate interest**, simple interest is calculated as a percentage of the initial value of an asset and is the same amount each time period.

straight-line depreciation *See* **flat rate depreciation**.

terms (of an arithmetic sequence) can represent the accumulation where simple interest is applied.

terms (of a geometric sequence) can represent the accumulation where compound interest is applied.

u_n The nth term of a sequence.

u_{n+1} The term after u_n.

unit cost depreciation The item depreciates through the amount it is used rather than its age.

useful life The time during which the item produces revenue for a business.

value of an asset As a recurrence relation for **flat rate depreciation** after the nth year is given by $u_0 = P$, $u_{n+1} = u_n - d$.

value of an asset As a recurrence relation, for **reducing balance depreciation** after the nth year is given by $u_0 = P$, $u_{n+1} = Ru_n$.

Exam practice

Multiple-choice questions

Appreciation and depreciation of assets: 17 questions

Solutions to this section start on page 246.

Question 1 ●●●

The recurrence relation $t_1 = 2$, $t_{n+1} = 1.5t_n + 2$ defines a sequence. The third term of the sequence will be:

A 3   **B** 5   **C** 7   **D** 9.5   **E** 45

Question 2 ●●●

If V is the value of an asset and $V_1 = \$120$ and $V_2 = \$135$, then the relationship between V_1 and V_2 is

A $V_2 = 1.012V_1$ **B** $V_2 = 1.1V_1$ **C** $V_2 = 1.1111V_1$

D $V_2 = 1.125V_1$ **E** $V_2 = 1.15V_1$

Question 3 ●●●

A recurrence relation is defined as $t_3 = 11$, $t_{n+1} = 2t_n - 5$.

The first term of this sequence is

A −1   **B** 6   **C** 6.5   **D** 8   **E** 29

Question 4 ●●●

Which one of the graphs below would represent the amount depreciated each year if a flat rate depreciation is applied?

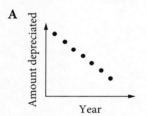
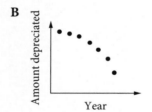
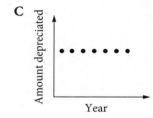
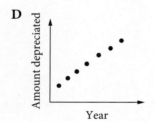
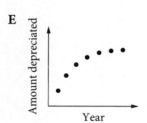

Question 5 ●●●

An asset is depreciating in value using flat-rate depreciation. If the purchase price of the asset is $4850 and its scrap value after five years is $1500, then the depreciation percentage is closest to

A 10%   **B** 11%   **C** 12%   **D** 14%   **E** 15%

Question 6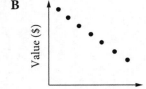

The values of an asset over three consecutive time periods are $1500, $1650, $1815.

Which one of the following is **true** for the three values of this asset?

A It represents a simple interest increase of 10%.

B They can be represented by the recurrence relation $V_0 = 1500$, $V_{n+1} = V_n + 150$.

C They can be represented by the recurrence relation $V_0 = 1500$, $V_{n+1} = 1.01V_n$.

D They represent a compound interest increase of 10%.

E They can be represented by the recurrence relation $V_0 = 1500$, $V_{n+1} = 0.1V_n$.

Question 7

Paul has purchased an equipment for $20 000 and will depreciate its value by 15% each year using the reducing-balance method. To determine the value, V_4, of the equipment after four years, Paul could use the recurrence relation

A $V_0 = 20\,000$, $V_{n+1} = 1.15V_n$ B $V_0 = 20\,000$, $V_{n+1} = 0.85V_n$ C $V_0 = 20\,000$, $V_{n+1} = (1.15)^4V_n$

D $V_0 = 20\,000$, $V_{n+1} = 1.85V_n$ E $V_0 = 20\,000$, $V_{n+1} = (0.85)^4V_n$

Question 8

An allowance was $232 in the year 2012. If the allowance has increased at a rate of 2% per year, then the allowance in the year 2016 was

A $246.20   B $250.56   C $251.12   D $256.55   E $315.63

Question 9

The recurrence relation $t_1 = 5$, $t_{n+1} = at_n - 2$ produces the sequence 5, 13, 37, 109 …

The value of a must be

A 2   B 3   C 5   D 6   E 9

Question 10

An item purchased at the beginning of 2014 depreciates by 15% each year using the reducing balance method. The first occasion when the value of the item is less than half the purchase price is

A at the beginning of 2017.   B at the beginning of 2018.   C at the beginning of 2019.

D at the beginning of 2020.   E at the beginning of 2021.

Question 11

The value of an item depreciates by 16% each year using the reducing balance method. Which one of the following graphs represents the value of the item over the first seven years?

A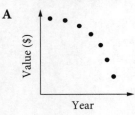

B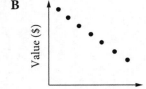

C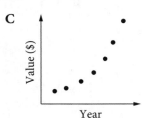

D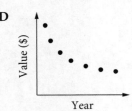

E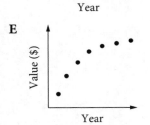

Question 12 ●●●

The book value of an asset is $4880 after five years. If the reducing balance depreciation rate is 15%, then the purchase price of the asset will be closest to

A $9815   **B** $10 540   **C** $11 000   **D** $19 520   **E** $28 060

Question 13 ●●●

A company van, bought for $32 000, is depreciated using the unit-cost method of depreciation. If the value of the van is $5000 after it has travelled 200 000 kilometres, then the rate of depreciation, in cents/kilometre, is

A 5   **B** 10   **C** 13.5   **D** 15   **E** 16

Question 14 ●●●

A photocopier is purchased for $16 450. The photocopier is depreciating 2.4 cents for each photocopy made. After four years of use, the book value of the photocopier is $9750.

The number of copies the photocopier has made in the four years is closest to

A 16   **B** 2800   **C** 70 000   **D** 160 800   **E** 280 000

Question 15 ●●●

Paul has purchased an equipment for $20 000 and will depreciate its value by 15% each year using the reducing balance method. To determine the value of the equipment after four years, Paul could use the expression

A $20\,000 \times 0.85 \times 4$ **B** $20\,000 \div 0.85^4$ **C** $20\,000 \times 0.85^4$

D $20\,000 - (4 \times 3000)$ **E** $20\,000 \times 1.15^4$

Question 16 ●●●

A computer with a purchase price of $4000 on 1 January 2014 is to be depreciated by 16% using the straight-line method. A recurrence relation that can be used to calculate the value V, of the computer, n years after 1 January 2014 is

A $V_0 = 4000, V_{n+1} = 0.84V_n$ **B** $V_0 = 4000, V_{n+1} = V_n - 0.16$

C $V_0 = 4000, V_{n+1} = 0.16V_n$ **D** $V_0 = 4000, V_{n+1} = V_n - 640$

E $V_0 = 4000, V_{n+1} = V_n + 640$

Question 17 ●●●

Margaret's laptop computer, originally costing $3000, has been depreciated in the last two years under the reducing balance method. If it is now worth $1920, then the depreciation rate is

A 12%   **B** 15%   **C** 18%   **D** 20%   **E** 21%

Simple interest and compound interest investments and loans: 12 questions

Solutions to this section start on page 247.

Question 18 ⬤◯◯

Which one of the following is **true** about a simple interest increase in the value of an asset?

A The value of the asset increases by an increasing amount each time period.

B If the increase for each time period is graphed, then the graph will be a series of discrete points in a straight line with a positive gradient.

C A recurrence relation for a simple interest increase in value will have the form $u_0 = P$, $u_{n+1} = bu_n$ where P is the initial amount invested and b is a constant.

D A recurrence relation for a simple interest increase in value will have the form $u_0 = P$, $u_{n+1} = b$, where P is the initial amount invested and b is a constant.

E If the value of the asset is graphed over a number of time periods, the graph will be a series of discrete points in a straight line with a positive gradient.

Question 19 ⬤◯◯

Tony correctly used the formula $A = PR^n$ to calculate the value of $5000 invested for five years at 6% per annum, compounding quarterly. The value of R that Tony used was

A 1.005   B 1.006   C 1.015   D 1.05   E 1.06

Question 20 ⬤◯◯

If $6500 is invested for 15 months at 7.5% simple interest per annum calculated monthly, then the amount of interest earned in this time is

A $50.78   B $73.13   C $609.38   D $731.25   E $812.50

Question 21 ⬤⬤◯

Anita has invested $12 000 for a period of three years. The interest rate for Anita's investment is 6.4% p.a., compounding quarterly. An expression for the value of Anita's investment, in dollars, at the end of the three-year term is

A $12\,000 \times 1.016^{12}$ B $12\,000 \times 1.016^{3}$ C $12\,000 \times 0.064^{12}$

D $12\,000 \times 1.064^{3}$ E $12\,000 \times 1.064$

Question 22 ⬤⬤◯

Jessica has invested $5000 in a term deposit account earning 5.5% p.a. interest for three months, with interest calculated at the end of the term. The amount she will collect after the three months will be

A $5068.75   B $5206.25   C $5275.00   D $6318.75   E $6875

Question 23 ⬤⬤◯

Brett borrows $280 000 for a period of 21 days as bridging finance for his new home. He is charged a rate of 7.75% p.a. for the loan, calculated when the loan is repaid. The amount of interest, to the nearest dollar, he will pay on this loan is

A $125   B $217   C $456   D $1248   E $2170

Question 24 ●●

An investment of $5000 is being credited with 5.2% p.a. interest that is compounded annually.
A recurrence relation to represent V_n, the value of the investment after n years is

A $V_0 = 5000$, $V_{n+1} = 5.2V_n$ **B** $V_0 = 5000$, $V_{n+1} = 0.052V_n$ **C** $V_0 = 5000$, $V_{n+1} = 1.52V_n$

D $V_0 = 5000$, $V_{n+1} = 1.052V_n$ **E** $V_0 = 5000$, $V_{n+1} = 1.0525V_n$

Question 25 ●●

$10\,000$ is invested in an account earning 6.8% p.a. interest compounding quarterly. Which one
of the following graphs best illustrates the amount of interest earned by this investment each quarter
for two years?

A 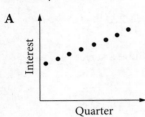   **B** 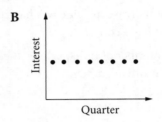   **C**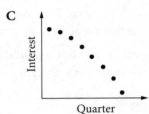

D 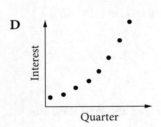   **E**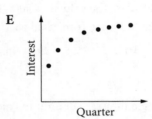

Question 26 ●●

An investment is being paid interest of 6% p.a. compounding yearly. The number of years before
the investment has doubled in value is

A 5   **B** 9   **C** 10   **D** 11   **E** 12

Question 27 ●●

The value of an investment increases by 3% each year compounding yearly. To calculate the value of the
investment in five years' time we would

A multiply the present value by 0.15.   **B** multiply the present value by 1.03^5.

C multiply the present value by $(1 + 0.03)^5$.   **D** divide the present value by $(1 - 0.03)^5$.

E divide the present value by 0.975.

The following information relates to Questions 28.

A loan of $5400 is being charged a flat rate of 10.5% p.a. interest over five years.

Question 28 ●●

If the principal and interest are being repaid in 60 equal monthly repayments, then the monthly
repayment will be

A $47.25   **B** $58.27   **C** $99.45   **D** $137.25   **E** $148.27

Question 29 ●●●

Interest is paid at 4.95% p.a. compounding monthly on an investment of $15\,000. The amount of interest
earned after three years is

A $186.39   **B** $2227.50   **C** $2339.58   **D** $2385.54   **E** $2396.08

Reducing balance loans: 11 questions

Solutions to this section start on page 248.

Question 30 ⚪⚫⚫

Rodger has borrowed $250 000 to buy a new home and will repay the loan with monthly repayments over 25 years. Interest is charged at 6.5% compounding monthly. The amount that Rodger needs to pay each month is closest to

A $833 **B** $1192 **C** $1402 **D** $1688 **E** $1807

Question 31 ⚪⚫⚫

Jim has a reducing balance loan of $312 700 that he is repaying at $2043.50 per month over a term of 25 years. The annual rate of interest that is being charged on this loan is

A 6% **B** 6.1% **C** 6.15% **D** 6.2% **E** 6.25%

Question 32 ⚫⚫⚫

The recurrence relation $P_0 = 456\,000$, $P_{n+1} = 1.005\,25P_n - 3022$ represents a reducing balance loan, where P_n is the amount owing after the nth month.

The yearly interest rate and the monthly repayment amount for this loan, respectively, are

A 5.25%, $3022 **B** 5.25%, $456 000 **C** 1.005 25%, $3022

D 6.3%, $456 000 **E** 6.3%, $3022

Question 33 ⚫⚫⚫

Jim and Alice have a reducing balance loan of $520 000 at 5.4% p.a. to be repaid over 25 years.

The difference in interest paid, in dollars, if they make monthly payments rather than quarterly payments is closest to

A 0 **B** 1564 **C** 2010

D 3487 **E** cannot be determined

Question 34 ⚫⚫⚫

Joan and Richard have a reducing balance loan on which they are making fortnightly repayments. If they change their payments from fortnightly to monthly, which one of the following will **not** be true?

A If their interest rate and term do not change, then the total amount they pay per year will increase.

B If their interest rate and term do not change, then the amount of interest that they pay over the term of the loan will increase.

C If their interest rate does not change and they pay the same total amount per year, then the term will increase.

D If their interest rate does not change, then they will pay the same amount of interest over the term of the loan.

E If their interest rate does not change, the term will increase unless they increase the total amount they pay each year.

CHAPTER 2 – EXAM PRACTICE

Question 35 ⬤⬤◯

The recurrence relation $V_0 = 425\,000$, $V_{n+1} = 1.0055V_n - 2096$ could represent

A an annuity, paying 6.6% p.a., where the initial investment is $425\,000 and the annuity pays 2096 per year.

B an annuity, paying 5.5% p.a., where the initial investment is $425\,000 and the annuity pays 2096 per month.

C a reducing balance loan of $425\,000 charging 6.6% p.a., where the repayments are $2096 per month.

D a reducing balance loan of $425\,000 charging 5.5% p.a., where the repayments are $2096 per month.

E an investment of $425\,000 at 6.6% p.a., with additional payments of $2096 per month.

Question 36 ⬤⬤◯

Bryan is repaying $615 per month on his loan of $210\,000. If he is being charged 6.65% p.a. on this loan, which one of the following graphs charts the amount still owing on Bryan's loan over a number of years?

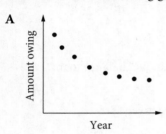
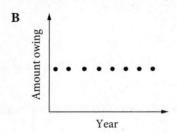
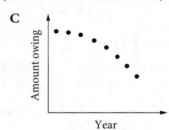

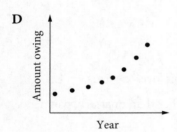
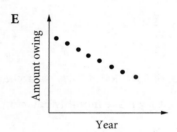

Question 37 ⬤⬤◯

A reducing balance loan is paid out over a term of eight years. Which one of the following graphs best represents the amount of interest paid each year?

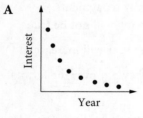
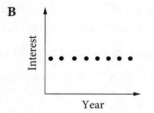
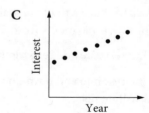

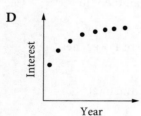
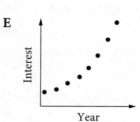

Question 38 ⬤⬤⬤

Ned is investing $10\,000 at a rate of 6.2% p.a. compounding monthly over 5 years. If the interest rate is changed to 7.2%, then the extra amount, to the nearest dollar, that Ned will have in his account after five years will be

A $42   **B** $666   **C** $673   **D** $695   **E** $107

Question 39 ●●●

Trevor has a personal loan of $18 000 and is being charged 7.2% p.a. interest, compounding monthly.
He is planning to repay $600 per month; the payment being made after interest is charged. The amount
that he will owe after his first payment is

A $17 400   **B** $17 600   **C** $17 529.60   **D** $17 504.40   **E** $17 508

Question 40 ●●●

The recurrence relation $P_{n+1} = 1.057P_n - 1250$ for a reducing balance loan gives the value $P_2 = 7205$,
where P_n is the amount owing after the nth year.

The initial amount borrowed P_0, in dollars, for this loan is closest to

A 8000   **B** 8750   **C** 8880   **D** 9180   **E** 9570

Annuities, perpetuities and annuity investments: 11 questions

Solutions to this section start on page 250.

Question 41 ●●○

Stefan has purchased a perpetuity with his superannuation that will pay him an income each month.
He is investing $560 000 at a rate of 4.9% p.a. for ten years.

The monthly amount that Stefan will receive is

A $2286.67   **B** $2862.76   **C** $3427.82   **D** $4472.85   **E** $5912.33

Question 42 ●●○

Ian has calculated that he needs $2200 per month for living expenses. The amount, to the nearest dollar,
that he will need to invest, at 6.75% p.a. interest, to provide this amount per month for an indefinite
period is

A $32 593   **B** $137 398   **C** $289 104   **D** $365 652   **E** $391 111

Question 43 ●●○

Kate has $174 000 to invest in a perpetuity. The amount she will receive each month, to the nearest dollar,
if she can invest her money at an interest rate of 7.35% p.a. is

A $1066   **B** $1078   **C** $1102   **D** $12 789   **E** $13 229

Question 44 ●●○

Bianca has started a savings account with an initial deposit of $2000. She is going to add $300 each month
to this account and it will be credited with 5.2% p.a. interest compounded monthly. The amount in the
account after two years will be

A $5351.76   **B** $9231.20   **C** $9789.17   **D** $10 206.03   **E** $10 987.33

Question 45 ●●○

Amy plans to have $40 000 saved as a deposit for a house in three years' time. She already has $12 000
saved in an account paying 5.5% p.a. interest, compounding monthly, and is planning to add to this each
month. The amount per month, to the nearest dollar, that Amy will need to add to her account so that
she has $40 000 in three years' time is closest to

A $662   **B** $684   **C** $718   **D** $778   **E** $1387

Question 46 ⚫⚫⚪

Chen has deposited $7500 in an account paying interest of 5.4% p.a. compounding fortnightly and he is going to add $150 to this account each fortnight. The amount, to the nearest dollar, in Chen's account after two years will be

A 11 413   **B** 11 919   **C** 12 556   **D** 16 582   **E** 18 238

Question 47 ⚫⚫⚫

Ellen has invested her savings of $8500 in an account paying 5.4% p.a. interest, compounding monthly. The first occasion when she will have more than $10 000 in her account will be

A after 3 years   **B** after 37 months   **C** after 38 months

D after 40 months   **E** after 41 months

Question 48 ⚫⚫⚪

The recurrence relation $A_0 = 480\,000$, $A_{n+1} = 1.003\,75A_n + 800$ could represent

A an annuity paying 3.75% p.a., where the initial investment is $480 000 and the annuity pays 800 per month.

B an annuity paying 4.5% p.a., where the initial investment is $480 000 and the annuity pays $800 per month.

C a reducing balance loan of $480 000 charging 4.5% p.a., where the repayments are $800 per month.

D a reducing balance loan of $480 000 charging 3.75% p.a., where the repayments are $800 per month.

E an investment of $480 000 at 4.5% p.a., with additional payments of $800 per month.

Question 49 ⚫⚫⚪

Liam has an account where he has saved $18 000 already. He wants to add monthly payments to this account to save at least $80 000 in three years' time.

If the interest rate is 5.55% p.a. compounding monthly, what is the minimum amount, to the nearest dollar, that Liam needs to save each month to achieve his aim?

A 1504   **B** 1723   **C** 1883   **D** 2592   **E** 3260

The following information relates to Questions 50 and 51.

Helene has an annuity investment with an initial value of $560 000. The investment has a return of 4.5% p.a. compounding monthly. Each month, Helene receives $3800.

Question 50 ⚫⚫⚪

A recurrence relation for this annuity is

A $A_0 = 560\,000$, $A_{n+1} = 1.0375A_n - 3800$ **B** $A_0 = 560\,000$, $A_{n+1} = 1.045A_n + 3800$

C $A_0 = 560\,000$, $A_{n+1} = 1.045A_n - 3800$ **D** $A_0 = 560\,000$, $A_{n+1} = 1.003\,75A_n - 3800$

E $A_0 = 560\,000$, $A_{n+1} = 1.0375A_n + 3800$

Question 51 ⚫⚫⚪

Helene's investment will last for approximately

A 12 years   **B** 18 years   **C** 21 years   **D** 24 years   **E** 25 years

Question 52 ©VCAA 2016S 1CQ18 ●○○

Which of the following recurrence relations will generate a sequence whose values decay geometrically?

A $L_0 = 2000$, $L_{n+1} = L_n - 100$ **B** $L_0 = 2000$, $L_{n+1} = L_n + 100$ **C** $L_0 = 2000$, $L_{n+1} = 0.65L_n$

D $L_0 = 2000$, $L_{n+1} = 6.5L_n$ **E** $L_0 = 2000$, $L_{n+1} = 0.85L_n - 100$

Question 53 ©VCAA 2016S 1CQ19 ●●○

Eva has $1200 that she plans to invest for one year.

One company offers to pay her interest at the rate of 6.75% per annum compounding daily.

The effective annual interest rate for this investment would be closest to

    A 6.75%   **B** 6.92%   **C** 6.96%   **D** 6.98%   **E** 6.99%

Use the following information to answer Questions 54, 55 and 56.

Kim invests $400 000 in an annuity paying 3.2% interest per annum.

The annuity is designed to give her an annual payment of $47 372 for 10 years.

The amortisation table for this annuity is shown below.

Some of the information is missing.

Payment number (n)	Payment	Interest earned	Reduction in principal	Balance of annuity
0	0.00	0.00	0.00	400 000.00
1	47 372.00	12 800.00	34 572.00	
2	47 372.00	11 693.70	35 678.30	329 749.70
3	47 372.00	10 551.99	36 820.01	292 929.69
4	47 372.00	9 373.75	37 998.25	254 931.44
5	47 372.00	8 157.81		215 717.24
6	47 372.00	6 902.95	40 469.05	175 248.19
7	47 372.00	5 607.94	41 764.06	133 484.14
8	47 372.00			90 383.63
9	47 372.00	2 892.28	44 479.72	45 903.90
10	47 372.00	1 468.92	45 903.08	0.83

Question 54 ©VCAA 2016S 1CQ21 ●○○

The balance of the annuity after one payment has been made is

A $339 828.00   **B** $352 628.00   **C** $365 428.00   **D** $387 200.00   **E** $400 000.00

Question 55 ©VCAA 2016S 1CQ22 ●○○

The reduction in the principal of the annuity after payment number 5 is

A $36 820.01   **B** $37 998.25   **C** $39 214.19   **D** $40 469.05   **E** $41 764.06

Question 56 ©VCAA 2016S 1CQ23 ●●○

The amount of payment number 8 that is the interest earned is closest to

A $3799.82   **B** $4074.67   **C** $4271.49   **D** $4836.57   **E** $5607.94

Question 57 ©VCAA 2016 1CQ19

The purchase price of a car was $26 000.

Using the reducing balance method, the value of the car is depreciated by 8% each year.

A recurrence relation that can be used to determine the value of the car after n years, C_n, is

A $C_0 = 26\,000, C_{n+1} = 0.92C_n$

B $C_0 = 26\,000, C_{n+1} = 1.08C_n$

C $C_0 = 26\,000, C_{n+1} = C_n + 8$

D $C_0 = 26\,000, C_{n+1} = C_n - 8$

E $C_0 = 26\,000, C_{n+1} = 0.92C_n - 8$

Question 58 ©VCAA 2016S 1CQ20

Rohan invests $15 000 at an annual interest rate of 9.6% compounding monthly.

Let V_n be the value of the investment after n months.

A recurrence relation that can be used to model this investment is

A $V_0 = 15\,000, V_{n+1} = 0.96V_n$

B $V_0 = 15\,000, V_{n+1} = 1.008V_n$

C $V_0 = 15\,000, V_{n+1} = 1.08V_n$

D $V_0 = 15\,000, V_{n+1} = 1.0096V_n$

E $V_0 = 15\,000, V_{n+1} = 1.096V_n$

Question 59 ©VCAA 2016 1CQ23

Sarah invests $5000 in a savings account that pays interest at the rate of 3.9% per annum compounding quarterly. At the end of each quarter, immediately after the interest has been paid, she adds $200 to her investment.

After two years, the value of her investment will be closest to

A $5805   **B** $6600   **C** $7004   **D** $7059   **E** $9285

Question 60 ©VCAA 2016S 1CQ24

The following graph shows the decreasing value of an asset over eight years.

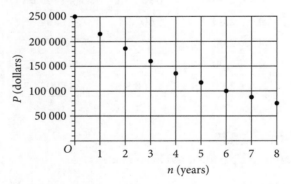

Let P_n be the value of the asset after n years, in dollars.

A rule for evaluating P_n could be

A $P_n = 250\,000 \times (1 + 0.14)^n$

B $P_n = 250\,000 \times 1.14 \times n$

C $P_n = 250\,000 \times (1 - 0.14) \times n$

D $P_n = 250\,000 \times (0.14)^n$

E $P_n = 250\,000 \times (1 - 0.14)^n$

Question 61 ©VCAA 2017N 1CQ23 ●●●

The value of a piano is depreciated using the reducing balance method.

The graph below shows the value of the piano as it depreciates over a period of 10 years.

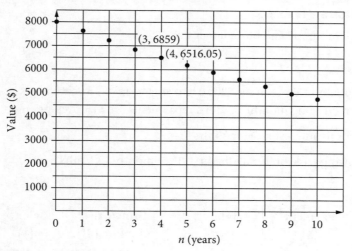

Let P_n be the value of the piano after n years.

A recurrence relation that could be used to determine P_n is

A $P_0 = 8000$, $P_{n+1} = 0.95P_n$

B $P_0 = 8000$, $P_{n+1} = 342.95P_n$

C $P_0 = 8000$, $P_{n+1} = 1.05P_n - 3$

D $P_0 = 8000$, $P_{n+1} = 0.95P_n - 342.95$

E $P_0 = 8000$, $P_{n+1} = P_n - 342.95$

Question 62 ©VCAA 2017N 1CQ24 ●●●

Geoff has a compound interest investment that earns interest compounding monthly.

The balance of Geoff's compound interest investment was $4418.80 after six months.

The balance of Geoff's compound interest investment was $4862.80 after two years.

The amount of money that Geoff initially invested is closest to

A $4000   **B** $4015   **C** $4280   **D** $4370   **E** $4715

Question 63 ©VCAA 2017N 1CQ17 ●○○

A sequence is generated by the recurrence relation below.

$A_0 = 2$, $A_{n+1} = 3A_n - 3$

The sequence is

A 2, 1, 0, –3…   **B** 2, 3, 0, –3…   **C** 2, 3, 2, 3…   **D** 2, 3, 3, 3…   **E** 2, 3, 6, 15…

Question 64 ©VCAA 2017N 1CQ21 ●●○

The amortisation table below shows the repayment, interest, principal reduction and balance of a reducing balance loan after the first repayment.

Repayment number	Repayment	Interest	Principal reduction	Balance of loan
0	0.00	0.00	0.00	180 000.00
1	850.00	720.00	130.00	179 870.00
2	850.00			

What amount of interest is paid with Repayment number 2?

A $608.56   **B** $609.44   **C** $717.12   **D** $719.48   **E** $720.00

Extended-answer questions

Solutions to this section start on page 252.

65 Changing houses (15 marks)

Vinh is planning to sell his home and buy a larger one.

Question 65.1 (12 marks)

Vinh will need to negotiate a new loan for $235 000 over a term of 25 years. He is being charged 6.3% p.a. interest, compounding monthly and his repayments are $1557.49 per month.

a Determine the monthly interest rate for Vinh's loan.   1 mark

b Complete the following table outlining the first three months of Vinh's loan and repayments.   3 marks

Payment number	Payment ($)	Interest ($)	Principal reduction ($)	Balance ($)
0	0.00	0.00	0.00	235 000.00
1	1557.49	1233.75	323.74	234 676.26
2	1557.49			
3	1557.49			

c A recurrence relation for Vinh's loan has the form

$$V_0 = 235\,000, \quad V_{n+1} = aV_n + b,$$

where V_n is the amount that Vinh owes after the nth month.

What are the values of a and b?   2 marks

d How much will Vinh pay in interest over the term of the loan? Give your answer correct to the nearest ten dollars.   2 marks

e After which month will he have repaid at least half of the original principal?   1 mark

f How much of his

  i first repayment   1 mark

  ii one hundredth repayment   2 marks

will go towards repaying the loan? Give your answers correct to the nearest dollar.

Question 65.2 (3 marks)

Vinh has the option of making 26 fortnightly repayments per year. The interest charged is 6.3% p.a., compounding fortnightly, over 25 years.

a How much will he need to repay each fortnight? Give your answer correct to the nearest cent.   1 mark

b How much will he save in interest if he makes fortnightly repayments, compared to the monthly repayments, over the term of the loan? Give your answer correct to the nearest ten dollars.   2 marks

66 Dave's Dilemma (14 marks)

Question 66.1 (3 marks) ▢▪▪

Dave is about to buy a sports car valued at $100 000. He will need to borrow the money in order to finance his purchase. The first option he looks at is as follows.

Option 1: Borrow the $100 000 from Duncan Inc. and repay the money in full in ten years with a flat interest rate of 6% per annum. Repayments are to be made monthly.

a Calculate the amount of interest Dave will pay each year under these terms.   1 mark

b Calculate the total of all payments that Dave will have made to Duncan Inc. once the loan has been entirely paid off.   1 mark

c What effect on total repayments will occur if Dave is to make fortnightly repayments rather than monthly?   1 mark

Question 66.2 (2 marks) ▢▪▪

Dave decides to look around at other alternatives. He considers option 2 as described below.

Option 2: His friend Fletch offers to lend him $100 000, and charges him $40 000 interest. Dave has to repay the total money in five years, with repayments to be made monthly.

a Calculate the annual flat rate of interest charged by Fletch.   1 mark

b Compared to Duncan Inc.'s terms, how much would Dave save by accepting the terms offered by Fletch?   1 mark

Question 66.3 (5 marks) ▢▪▪

Another alternative for Dave is to borrow an additional $100 000 from his bank. He already has a mortgage of $60 000.

Option 3: Dave will repay the $160 000 owing over 20 years at 7.5% per annum on the reducing monthly balance. Repayments will be made monthly.

a Determine the monthly repayments that Dave must make in order to pay off the loan in the specified time. Give your answer to the nearest dollar.   1 mark

b A recurrence relation for Dave's loan has the form $D_0 = 160\,000$, $D_{n+1} = aD_n + b$, where D_n is the amount Dave owes after the nth month.

Determine the values of the constants a and b.   2 marks

c Dave decides he could actually repay $1500 per month instead of the figure found in part **a**. Determine the overall saving he would make by paying this extra amount. Give your answer to the nearest ten dollars.   2 marks

Question 66.4 (4 marks)

Dave's sports car is depreciating in value by 15% each year.

a Determine the value of his sports car at the end of the first year.   1 mark

b Determine the book value of Dave's sports car after five years. Give your answer to the nearest dollar.   1 mark

c If Dave borrowed the money to buy the car from his friend Fletch (Option 2) and he sold the car for the book value after five years, how much in total has it cost Dave to own the car for five years? Give your answer to the nearest dollar.   2 marks

67 Living it up (15 marks)

Sergio has \$780 000 to provide for his retirement.

Question 67.1 (3 marks)

Sergio can invest his money in a perpetuity that will provide a monthly payment
for as long as he lives.

a How much will he receive each month if his money is invested at 4.5% p.a.?   2 marks

b If Sergio lives for another 15 years, how much of the original investment will be left to go
to his estate?   1 mark

Question 67.2 (5 marks)

Sergio estimates that he will need \$4000 per month to live comfortably in his retirement.
He has found an investment that pays 4.5% p.a. compounding monthly.

a After one monthly payment of \$4000, how much will he have left of his \$780 000?   1 mark

b A recurrence relation for S_n, the amount of Sergio's investment after n months,
has the form $S_0 = 780\,000$, $S_{n+1} = aS_n + b$.

Determine the value of the constants a and b.   2 marks

c After 10 years of monthly payments of \$4000, how much will he have left of his
original investment?   1 mark

d How long will his investment last if he receives monthly payments of \$4000?
Give your answer to the nearest month.   1 mark

Question 67.3 (2 marks)

If inflation averages 2.5% per year, how much, to the nearest dollar, will Sergio need
in ten years' time so that he has the same buying power as \$4000 today?

Question 67.4 (3 marks)

One option suggested to Sergio is to provide \$4000 per month for the first five years and
then \$5000 per month for the remainder of the time that his investment lasts. His investment
of \$780 000 earns an interest rate of 4.5% p.a.

a How much will he have left of the original amount after five years? Give your answer
to the nearest cent.   2 marks

b How long will the investment last overall? Give your answer to the nearest month.   1 mark

Question 67.5 (2 marks)

Sergio decides that he wants his money to last 15 years, after which time there will not be
any of the original investment left.

If his investment of \$780 000 earns an interest rate of 4.8% p.a., how much will he receive
per month? Give your answer to the nearest dollar.

68 Wheeler dealing (15 marks)

Jake is planning to acquire a new car and the price of his chosen vehicle is $34 600.

Question 68.1 (1 mark)

If Jake leases the car, he will be required to make payments of $665 per month. The leasing contract will continue for three years, after which time the value of the car will be assumed to be $18 000. He will have the option of buying the car after the three years for $18 000.

How much will Jake pay in total for the lease of the car over the three-year period?

Question 68.2 (2 marks)

If Jake had borrowed the purchase price of the car, paid $665 per month on the reducing amount owed, with $18 000 still owing after three years, what annual rate of interest would he be charged? Give your answer correct to one decimal place.

Question 68.3 (5 marks)

Another option available to Jake is to make a $3460 deposit plus monthly payments of $732 over five years.

a Show that the total amount paid in interest under this plan is $12 780.   1 mark

b If the interest is quoted at a flat rate, calculate the annual flat rate of interest being
        charged. Give your answer correct to one decimal place.   2 marks

c Calculate the equivalent reducing balance rate of interest for this plan. Give your
        answer correct to two decimal places.   2 marks

Question 68.4 (7 marks)

If the value of the car depreciates over the three years from $34 600 to $18 000,

a at a flat rate:

    i calculate the annual flat rate (to two decimal places)   1 mark

    ii write a recurrence relation for the value, V_n, of the car after n years in the form
      $V_0 = 34\,600$, $V_{n+1} = aV_n + b$, where a and b are constants   1 mark

b at a reducing balance rate:

    i calculate the annual rate (to two decimal places)   1 mark

    ii write a recurrence relation for the value, R_n, of the car after n years in the form
      $R_0 = 34\,600$, $R_{n+1} = aR_n + b$;, where a and b are constants   2 marks

c at a unit cost rate, calculate the rate, in cents per kilometre, given that the car has travelled
        75 000 km in three years.   2 marks

69 Original concepts (15 marks)

Question 69.1 (9 marks)

Chris, the founder of the company Original Concepts Unlimited, has a state-of-the-art video camera, which was originally purchased for $6000. He plans to depreciate the camera over five years. The scrap value of the camera at the end of the five-year period is expected to be $600.

a Determine the amount by which Chris' camera will depreciate over the five years. 1 mark

b If Chris uses flat rate depreciation, determine the annual percentage rate of depreciation that he is using. 2 marks

c What percentage of the camera's value has been written off by Chris after five years? 1 mark

d Chris estimates that the camera's value depreciates, on average, by $4 for each hour of use. How many hours, on average, does Chris use his camera each year? 1 mark

e If Chris depreciated the value of the camera to the same scrap value after five years using the reducing balance method, what is the annual depreciation rate? Give your answer correct to one decimal place. 2 marks

f Explain why the reducing balance depreciation rate found in part **e** is higher than the flat rate depreciation rate found in part **b**. 1 mark

g Describe the difference between the graphs of value versus time for these two types of depreciation. 1 mark

Question 69.2 (6 marks)

Chris decides to purchase new premises from which he will run his business. He borrows $160 000 over 25 years at 6.6% per annum on the reducing quarterly balance.

a Show that the quarterly interest rate is 1.65%. 1 mark

b Chris' repayments are $3278.09 per quarter.

Complete the following table outlining the first three quarters of Chris' loan and repayments. Give all values correct to the nearest cent. 3 marks

Payment number	Payment ($)	Interest ($)	Principal reduction ($)	Balance ($)
0	0.00	0.00	0.00	160 000.00
1	3278.09	2640.00	638.09	159 361.91
2	3278.09			
3	3278.09			

c Chris decides to investigate an interest-only loan. The interest rate is 6.8% p.a. over 25 years.

i How much will Chris' quarterly repayments be for this interest-only loan? 1 mark

ii How much will Chris owe on this loan after 10 years? 1 mark

70 Travelling (15 marks)

Rebecca is saving for an overseas trip she intends to take in three years' time. She estimates that she will need a total of $20 000 for this trip. At present, she has $7500 invested in an account that is being credited with 5.4% p.a. interest, compounding monthly. Give all answers correct to the nearest cent.

Question 70.1 (8 marks)

a How much will be in this account after three years if Rebecca does not make any further payments?

1 mark

b If Rebecca makes additional payments of $100 each month,

   i write down the values of the constants a and b in the recurrence relation $R_0 = 7500$, $R_{n+1} = aR_n + b$, where R_n is the amount in Rebecca's account at the end of the nth month

2 marks

   ii use your recurrence relation to find the amount in the account at the end of the third month

2 marks

   iii use CAS to determine the amount that Rebecca will have in her account at the end of three years.

1 mark

c How much will Rebecca need to pay per month into the account so that she has accumulated a total of $20 000 after three years?

1 mark

d How much will Rebecca need to pay per month into the account so that she has accumulated a total of $20 000 after three years, if the interest is credited daily (365 days per year)?

1 mark

Question 70.2 (5 marks)

Rebecca decides to invest $300 per month into her present account with a starting balance of $7500 and 5.4% p.a. interest credited monthly.

a How much will be in the account after 12 months?

1 mark

After 12 months, the interest rate increases to 6.4% p.a.

b In which month, from the start, will she now meet her target of $20 000?

2 marks

c What will be the total amount in her account after three years?

1 mark

d How much interest will she have earned in the three years?

1 mark

Question 70.3 (2 marks)

Rebecca finds that she has inherited an amount of money. She still wants to have a total of $20 000 in her overseas trip account in three years' time. How much will she need to add to the $7500 in the account now, as a single payment, so that it will be a total of $20 000 in three years' time? The account is credited with 5.4% interest, compounding monthly.

71 School funding (15 marks)

Question 71.1 (5 marks)

When Adam was born, his parents started a savings account for him with a deposit of $1000.

If the account pays 4.8% p.a. interest compounding monthly, and no further payments are made:

a	calculate the monthly rate of interest	1 mark
b	calculate the amount in the account after each of the first three months	2 marks
c	write down a recurrence relation of the form $u_0 = a$, $u_{n+1} = bu_n + c$, where a, b and c are constants and u_n represents the amount in the account after n months.	2 marks

Question 71.2 (5 marks)

After Adam's first year, his parents decide to add $100 each month to the account. Assume that there is $1050 in the account after Adam's first year and that the account pays 4.8% p.a. compounding monthly.

For this new savings plan,

a	calculate the amount in the account after the first month	1 mark
b	write down the values of the constants a and b in the recurrence relation $A_0 = 1050$, $A_{n+1} = aA_n + b$, where A_n is the amount in Adam's account at the end of the nth month	2 marks
c	use your expression from part **b** to calculate A_4	1 mark
d	use CAS to find A_{48}, giving your answer correct to the nearest cent.	1 mark

Question 71.3 (1 mark)

Adam's parents realise that they will need to add more to the account each month if it is going to pay for his secondary education.

Assuming that there is $6552 in the account after five years and the account is still paying 4.8% p.a. compounding monthly, find the amount, to the nearest cent, that they will need to deposit each month so that they have a total of $60 000 in the account after another 7 years.

Question 71.4 (2 marks)

The fees for the first year of secondary education at the school Adam is going to attend increase by an average of 2.5% each year.

If the fees are $10 000 when Adam is five, what will the fees be in seven years' time? Give your answer to the nearest dollar.

Question 71.5 (2 marks)

When Adam starts secondary school, his parents do not add any more to the account but use it to pay an annual amount, at the beginning of each of the six years, to cover his fees. The account still earns 4.8% p.a. on the balance, compounding monthly, and there is nothing left after the last payment.

Find the annual amount that is paid at the beginning of each year.

72 Martine's car (15 marks)

Question 72.1 (6 marks)

Martine's parents are going to lend their daughter $20 000 to buy a car. They are charging 4.8% p.a. interest, compounding yearly, and she is to repay the loan in five yearly equal payments.

a Determine the yearly amount that she should repay to ensure that the loan and interest is repaid after five years.   1 mark

b Martine is going to repay $4000 for the first four years, then the remainder at the end of the fifth year.

Complete the table below, detailing the interest and balance of the loan over the five years.   3 marks

Payment number	Payment ($)	Interest ($)	Principal reduction ($)	Balance ($)
0	0	0	0	20 000
1	4000	960	3040	16 960
2	4000			
3	4000			
4	4000			
5				

c If Martine's final payment is the total amount owing at the end of the fifth year, find the value of this final payment.   1 mark

d How much interest has Martine paid her parents over the five years?   1 mark

Question 72.2 (3 marks)

The value of Martine's car has depreciated over the five years that she has owned it.

a The car depreciated by 24% in the first year. If it was bought for $19 850, show that its value after one year is $15 086.   1 mark

b The car depreciated by 15% for the next four years. Calculate the value of the car after five years. Give your answer to the nearest dollar.   2 marks

Question 72.3 (4 marks)

After five years, Martine trades in her car on a new model. She is given a $7000 trade-in on a new car that has a purchase price of $24 620. The car dealer offers her a loan at a nominal rate of 7.5% p.a. interest and requiring 36 monthly payments of $600.

a State the amount that Martine will be borrowing from the car dealer.   1 mark

b How much interest is Martine paying on this loan over the three years?   1 mark

c Is Martine's loan a flat rate loan or a reducing balance loan? Give a reason for your answer.   2 marks

73 VCAA 2016 Exam 2 (12 marks)

Question 73.1 (5 marks) ©VCAA 2016 2CQ5 ●●

Ken has opened a savings account to save money to buy a new caravan.

The amount of money in the savings account after n years, V_n, can be modelled by the recurrence relation shown below.

$$V_0 = 15\,000, \quad V_{n+1} = 1.04 \times V_n$$

a How much money did Ken initially deposit into the savings account?   1 mark

b Use recursion to write down calculations that show that the amount of money in Ken's savings account after two years, V_2, will be $16\,224.   1 mark

c What is the annual percentage compound interest rate for this savings account?   1 mark

d The amount of money in the account after n years, V_n, can also be determined using a rule.

   i Complete the rule below by writing the appropriate numbers in the boxes provided.   1 mark

$$V_n = \boxed{}^{\,n} \times \boxed{}$$

   ii How much money will be in Ken's savings account after 10 years?   1 mark

Question 73.2 (3 marks) ©VCAA 2016 2CQ6 ●●●

Ken's first caravan had a purchase price of $38\,000.

After eight years, the value of the caravan was $16\,000.

a Show that the average depreciation in the value of the caravan per year was $2750.   1 mark

b Let C_n be the value of the caravan n years after it was purchased.

Assume that the value of the caravan has been depreciated using the **flat rate** method of depreciation.

Write down a recurrence relation, in terms of C_{n+1} and C_n, that models the value of the caravan.   1 mark

c The caravan has travelled an average of 5000 km in each of the eight years since it was purchased. Assume that the value of the caravan has been depreciated using the **unit cost** method of depreciation.

By how much is the value of the caravan reduced per kilometre travelled?   1 mark

Question 73.3 (4 marks) ©VCAA 2016 2CQ7 ●●●

Ken has borrowed $70\,000 to buy a new caravan.

He will be charged interest at the rate of 6.9% per annum, compounding monthly.

a For the first year (12 months), Ken will make monthly repayments of $800.

   i Find the amount that Ken will owe on his loan after he has made 12 repayments.   1 mark

   ii What is the total interest that Ken will have paid after 12 repayments?   1 mark

b After three years, Ken will make a lump sum payment of L in order to reduce the balance of his loan.

This lump sum payment will ensure that Ken's loan is fully repaid in a further three years.

Ken's repayment amount remains at $800 per month and the interest rate remains at 6.9% per annum, compounding monthly.

What is the value of Ken's lump sum payment, L? Round your answer to the nearest dollar.   2 marks

74 VCAA 2017 Exam 2 (12 marks)

Question 74.1 (5 marks) ©VCAA 2017 2CQ5 ○◐◑

Alex is a mobile mechanic. He uses a van to travel to his customers to repair their cars. The value of Alex's van is depreciated using the flat rate method of depreciation.

The value of the van, in dollars, after n years, V_n, can be modelled by the recurrence relation shown below.

$$V_0 = 75\,000, \quad V_{n+1} = V_n - 3375$$

a Recursion can be used to calculate the value of the van after two years.

Complete the calculations below by writing the appropriate numbers in the boxes provided.   2 marks

$V_0 = 75\,000$

$V_1 = 75\,000 - \boxed{} = 71\,625$

$V_2 = \boxed{} - \boxed{} = \boxed{}$

b   i By how many dollars is the value of the van depreciated each year?   1 mark

   ii Calculate the annual flat rate of depreciation in the value of the van. Write your answer as a percentage.   1 mark

c The value of Alex's van could also be depreciated using the reducing balance method of depreciation.

The value of the van, in dollars, after n years, R_n, can be modelled by the recurrence relation shown below.

$$R_0 = 75\,000, \quad R_{n+1} = 0.943R_n$$

At what annual percentage rate is the value of the van depreciated each year?   1 mark

Question 74.2 (4 marks) ©VCAA 2017 2CQ6 ●●○

Alex sends a bill to his customers after repairs are completed. If a customer does not pay the bill by the due date, interest is charged. Alex charges interest after the due date at the rate of 1.5% per month on the amount of the unpaid bill. The interest on this amount will compound monthly.

a Alex sent Marcus a bill of $200 for repairs to his car. Marcus paid the full amount one month after the due date. How much did Marcus pay?   1 mark

Alex sent Lily a bill for $428 for repairs to her car. Lily did not pay the bill by the due date.

Let A_n be the amount of this bill n months after the due date.

b Write down a recurrence relation, in terms of A_0, A_{n+1} and A_n, that models the amount of the bill.   2 marks

c Lily paid the full amount of her bill four months after the due date. How much interest was Lily charged? Round your answer to the nearest cent.   1 mark

Question 74.3 (3 marks) `©VCAA 2017 2CQ7` ●●●

Alex sold his mechanics' business for $360 000 and invested this amount in a perpetuity. The perpetuity earns interest at the rate of 5.2% per annum. Interest is calculated and paid monthly.

a What monthly payment will Alex receive from this investment?   1 mark

b Later, Alex converts the perpetuity to an annuity investment. This annuity investment earns interest at the rate of 3.8% per annum, compounding monthly.

For the first four years Alex makes a further payment each month of $500 to his investment. This monthly payment is made immediately after the interest is added.

After four years of these regular payments, Alex increases the monthly payment. This new monthly payment gives Alex a balance of $500 000 in his annuity after a further two years.

What is the value of Alex's new monthly payment? Round your answer to the nearest cent.   2 marks

75 VCAA 2017 NHT Exam 2 (12 marks)

Question 75.1 (5 marks) `©VCAA 2017N 2CQ5`

The snooker table at a community centre was purchased for $3000.

After purchase, the value of the snooker table was depreciated using the flat rate method of depreciation.

The value of the snooker table, V_n, after n years, can be determined using the recurrence relation below.

$$V_0 = 3000, \; V_{n+1} = V_n - 180$$

a What is the annual depreciation in the value of the snooker table?   1 mark

b Use recursion to show that the value of the snooker table after two years, V_2, is $2640.   1 mark

c After how many years will the value of the snooker table first fall below $2000?   1 mark

d The value of the snooker table could also be depreciated using the reducing balance method of depreciation.

After one year, the value of the snooker table is $2760.

After two years, the value of the snooker table is $2539.20.

  i Show that the annual rate of depreciation in the value of the snooker table is 8%.   1 mark

  ii Let S_n be the value of the snooker table after n years.

    Write down the recurrence relation, in terms of S_{n+1} and S_n, that can be used to determine the value of the snooker table after n years using the reducing balance method.   1 mark

Question 75.2 (4 marks) `©VCAA 2017N 2CQ6`

The community centre opened a savings account with Bank P.

Let P_n be the balance of the savings account n years after it was opened.

The value of P_n can be determined using the recurrence relation model below.

$$P_0 = A, \quad P_{n+1} = 1.056 \times P_n$$

The balance of the savings account one year after it was opened was $1584.

a Show that the value of A is $1500.   1 mark

b Write down the balance of the savings account four years after it was opened.   1 mark

c The balance of the savings account six years after it was opened was $2080.05. This $2080.05 was transferred into a savings account with Bank Q. This savings account pays interest at the rate of 5.52% per annum, compounding monthly.

Let Q_n be the balance of this savings account n months after it was opened. The value of Q_n can be determined from a rule.

Complete this rule by writing the missing values in the boxes provided below.   2 marks

$Q_n =$  $\times$ ^{n}

Question 75.3 (3 marks) ©VCAA 2017N 2CQ7

The community centre has received a donation of $5000. The donation is deposited into another savings account. This savings account pays interest compounding monthly.

Immediately after the interest has been added each month, the community centre deposits a further $100 into the savings account.

After five years, the community centre would like to have a total of $14 000 in the savings account.

a What is the annual interest rate, compounding monthly, that is required to achieve this goal? Write your answer correct to two decimal places.   1 mark

b The interest rate for this savings account is actually 6.2% per annum, compounding monthly. After 36 deposits, the community centre stopped making the additional monthly deposits of $100.

How much money will be in the savings account five years after it was opened?   2 marks

76 VCAA 2018 NHT Exam 2 (12 marks)

Question 76.1 (4 marks) ©VCAA 2018N 2CQ7

Roslyn invested some money in a savings account that earns interest compounding annually. The interest is calculated and paid at the end of each year.

Let V_n be the amount of money in Roslyn's savings account, in dollars, after n years.

The recursive calculations below show the amount of money in Roslyn's savings account after one year and after two years.

$V_0 = 5000$

$V_1 = 1.05 \times 5000 = 5250$

$V_2 = 1.05 \times 5250 = 5512.50$

a How much money did Roslyn initially invest?   1 mark

b How much interest in total did she earn by the end of the second year?   1 mark

c Let V_n be the amount of money in Roslyn's savings account, in dollars, after n years.

Write down the recurrence relation, in terms of V_0, V_{n+1} and V_n, that can be used to model the amount of money, in dollars, in Roslyn's savings account.   1 mark

d Roslyn plans to use her savings to pay for a holiday. The holiday will cost $6000.

What minimum annual percentage interest rate would have been required for Roslyn to have saved this $6000 after two years? Round your answer to one decimal place.   1 mark

Question 76.2 (5 marks)   ©VCAA   2018N 2CQ8

Richard will join Roslyn on the holiday. He will sell his stereo system to help pay for his holiday. The stereo system was originally purchased for $8500. Richard will sell the stereo system at a depreciated value.

a Richard could use a flat rate depreciation method.

Let S_n be the value, in dollars, of Richard's stereo system n years after it was purchased.

The value of the stereo system, S_n, can be modelled by the recurrence relation below.

$$S_0 = 8500, \quad S_{n+1} = S_n - 867$$

   i Using this depreciation method, what is the value of the stereo system four years after it was purchased?   1 mark

   ii Calculate the annual percentage flat rate of depreciation for this depreciation method.   1 mark

b Richard could also use a reducing balance depreciation method, with an annual depreciation rate of 8%.

Using this depreciation method, what is the value of the stereo system four years after it was purchased? Round your answer to the nearest cent.   1 mark

c Four years after it was purchased, Richard sold his stereo system for $4500.

Assuming a reducing balance depreciation method was used, what annual percentage rate of depreciation did this represent? Round your answer to one decimal place.   2 marks

Question 76.3 (3 marks)   ©VCAA   2018N 2CQ9

Andrew will also join Roslyn and Richard on the holiday.

Andrew borrowed $10 000 to pay for the holiday and for other expenses.

Interest on this loan will be charged at the rate of 12.9% per annum, compounding monthly.

Immediately after the interest has been calculated and charged each month, Andrew will make a repayment.

a For the first year of this loan, Andrew will make interest-only repayments each month.

What is the value of each interest-only repayment?   1 mark

b For the next three years of this loan, Andrew will make equal monthly repayments. After these three years, the balance of Andrew's loan will be $3776.15.

What amount, in dollars, will Andrew repay each month during these three years?   1 mark

c Andrew will fully repay the outstanding balance of $3776.15 with a further 12 monthly repayments.

The first 11 repayments will each be $330.

The twelfth repayment will have a different value to ensure the loan is repaid exactly to the nearest cent.

What is the value of the twelfth repayment? Round your answer to the nearest cent.   1 mark

UNIT 4

Chapter 3 Matrices
Area of Study 2: Discrete mathematics

Content summary notes

Matrices

Matrices and their applications

- Definition of a matrix
- Elements of matrices
- Types of matrices
- Transpose of a matrix
- Matrix operations
- Inverse matrices
- Binary matrices
- Equality of matrices

Transition matrices

- Matrix recurrence relations
- Leslie matrices

Area of Study 2 Outcome 1

These are the key knowledge points for this chapter, please note that not all will be examinable.

- the order of a matrix, types of matrices (row, column, square, diagonal, symmetric, triangular, zero, binary, permutation and identity), the transpose of a matrix, and elementary matrix operations (sum, difference, multiplication of a scalar, product and power)

- the inverse of a matrix and the condition for a matrix to have an inverse, including determinant for transition matrices, assuming the next state only relies on the current state with a fixed population

- communication and dominance matrices and their application

- transition diagrams and transition matrices and regular transition matrices and their identification

Although you should become familiar with all of the key skills not all the skills are required for the exam. The key skills for Unit 4 Area of Study 2 are:

- use matrix recurrence relations to generate a sequence of state matrices, including an informal identification of the equilibrium or steady state matrix in the case of regular state matrices

- construct a transition matrix from a transition diagram or a written description and vice versa

- construct a transition matrix to model the transitions in a population with an equilibrium state

- use matrix recurrence relations to model populations with culling and restocking

VCE Mathematics Study Design 2023–2027 p. 91, © VCAA 2022

3.1 Matrices and their applications

3.1.1 Definition of a matrix

A **matrix** is a rectangular array of numbers usually denoted by capital letters $A, B, C \ldots$

The numbers in a matrix are usually referred to as **elements**.

A matrix of **order $m \times n$ has m rows and n columns**.

For example, $A = \begin{bmatrix} 1 & 4 & -4 \\ 6 & -2 & 0 \end{bmatrix}$ is a 2 × 3 matrix.

3.1.2 Elements of matrices

The element in the ith row and jth column of a matrix is denoted by a_{ij}.

In the matrix A above, the element in row one, column three, is −4. This can be abbreviated as $a_{13} = -4$.

3.1.3 Types of matrices

A matrix of order $1 \times n$ has one row and is called a **row matrix** or **row vector**.

For example, $\begin{bmatrix} 3 & 4 & 8 \end{bmatrix}$ is a 1 × 3 row matrix.

A matrix of order $m \times 1$ has m rows and one column and is called a **column matrix** or **column vector**.

For example, $\begin{bmatrix} 1 \\ 2 \\ 3 \\ 4 \end{bmatrix}$ is a 4 × 1 column matrix.

A **square matrix** has **order $m \times m$**. It has the same number of rows as columns.

For example, $\begin{bmatrix} 2 & 3 \\ -1 & 4 \end{bmatrix}$ is a 2 × 2 square matrix.

A **diagonal matrix** is a square matrix where all the elements, except those along the leading diagonal are zero.

For example, $\begin{bmatrix} 5 & 0 \\ 0 & -2 \end{bmatrix}$, $\begin{bmatrix} 1 & 0 & 0 \\ 0 & -3 & 0 \\ 0 & 0 & 2 \end{bmatrix}$

A **symmetric matrix** is a square matrix that is equal to its transpose.

For example, $\begin{bmatrix} 3 & 1 \\ 1 & 3 \end{bmatrix}^T = \begin{bmatrix} 3 & 1 \\ 1 & 3 \end{bmatrix}$

A **triangular matrix** is a square matrix where all the elements above, or below, the main diagonal are zero.

For example, $\begin{bmatrix} 1 & 0 & 0 \\ 2 & 1 & 0 \\ 3 & 2 & 1 \end{bmatrix}$ (lower triangular matrix) $\begin{bmatrix} 4 & 5 & 6 \\ 0 & 4 & 5 \\ 0 & 0 & 4 \end{bmatrix}$ (upper triangular matrix)

Every element of a **zero matrix** is zero. The zero matrix is denoted by 0. If a matrix is multiplied by a suitable zero matrix, then the result will be a zero matrix, i.e., $A \times 0 = 0 \times A = 0$.

3.1.3.1 Negative matrices

The negative of matrix A is the matrix $-A$.

$-A = -1 \times A$

$-A$ is found by changing the sign of each element in matrix A.

$A + (-A) = 0$

Example 1

If A is the matrix $\begin{bmatrix} 1 & 0 \\ -2 & 3 \end{bmatrix}$, then $-A$ is the matrix $\begin{bmatrix} -1 & 0 \\ 2 & -3 \end{bmatrix}$.

3.1.3.2 Identity matrix

An **identity matrix**, I, is a square matrix such that $A \times I = I \times A = A$.

An identity matrix has **elements of 1 along the leading diagonal and 0 for the other elements**.

The 1×1 identity matrix, I_1, is $\begin{bmatrix} 1 \end{bmatrix}$, the 2×2 identity matrix, I_2, is $\begin{bmatrix} 1 & 0 \\ 0 & 1 \end{bmatrix}$, the 3×3 identity matrix,

I_3, is $\begin{bmatrix} 1 & 0 & 0 \\ 0 & 1 & 0 \\ 0 & 0 & 1 \end{bmatrix}$, etc.

3.1.4 Transpose of a matrix

The **transpose of a matrix** is a matrix where the rows and columns are interchanged: the first row becomes the first column, the second row becomes the second column, etc. The transpose of a matrix is denoted by A^T and can be found on CAS as an inbuilt function. If A is an $m \times n$ matrix, then A^T will be an $n \times m$ matrix.

For example, if $C = \begin{bmatrix} 2 & 2 & 2 \\ 3 & 3 & 3 \end{bmatrix}$, then $C^T = \begin{bmatrix} 2 & 3 \\ 2 & 3 \\ 2 & 3 \end{bmatrix}$.

3.1.5 Matrix operations

3.1.5.1 Addition and subtraction

Addition and subtraction can only be carried out on matrices that have the **same order**. To add (or subtract), you add (or subtract) the corresponding elements.

Example 2

Adding two 2×2 matrices.

$$\begin{bmatrix} 3 & -1 \\ 4 & 0 \end{bmatrix} + \begin{bmatrix} 1 & 2 \\ 7 & -2 \end{bmatrix} = \begin{bmatrix} 3+1 & -1+2 \\ 4+7 & 0+(-2) \end{bmatrix}$$

$$= \begin{bmatrix} 4 & 1 \\ 11 & -2 \end{bmatrix}$$

Example 3

Subtracting two 2×1 matrices.

$$\begin{bmatrix} 4 \\ -2 \end{bmatrix} - \begin{bmatrix} 3 \\ 6 \end{bmatrix} = \begin{bmatrix} 4-3 \\ -2-6 \end{bmatrix} = \begin{bmatrix} 1 \\ -8 \end{bmatrix}$$

3.1.5.2 Scalar multiplication of matrices

If a matrix is multiplied by a scalar (a number) k, then every element in the matrix is multiplied by k.

Example 4

$$3 \times \begin{bmatrix} 2 & 6 & -2 \\ 3 & 1 & 0 \end{bmatrix} = \begin{bmatrix} 3 \times 2 & 3 \times 6 & 3 \times (-2) \\ 3 \times 3 & 3 \times 1 & 3 \times 0 \end{bmatrix}$$

$$= \begin{bmatrix} 6 & 18 & -6 \\ 9 & 3 & 0 \end{bmatrix}$$

3.1.5.3 Product of matrices

The product, AB, of two matrices A and B can only be found if the number of columns in A is equal to the number of rows in B.

Example 5

$$\begin{bmatrix} 2 & 1 & 3 \end{bmatrix} \times \begin{bmatrix} 1 & -2 \\ 0 & 4 \\ -5 & 3 \end{bmatrix}$$ can be found because it is the product of a 1×3 matrix and a 3×2 matrix.

The resulting matrix will be a 1×2 matrix.

$$\text{Same}$$
$$(1 \times 3) \times (3 \times 2) = (1 \times 2)$$
$$\text{Product}$$

$$\begin{bmatrix} 2 & 1 & 3 \end{bmatrix} \times \begin{bmatrix} 1 & -2 \\ 0 & 4 \\ -5 & 3 \end{bmatrix}$$

$$= \begin{bmatrix} \text{row 1} \times \text{column 1} & \text{row 1} \times \text{column 2} \end{bmatrix}$$

$$= \begin{bmatrix} (2 \times 1) + (1 \times 0) + 3 \times (-5) & 2 \times (-2) + (1 \times 4) + (3 \times 3) \end{bmatrix}$$

$$= \begin{bmatrix} -13 & 9 \end{bmatrix}$$

The product $\begin{bmatrix} 3 & 2 & -1 \\ 0 & 6 & 8 \end{bmatrix} \times \begin{bmatrix} -2 & 1 \end{bmatrix}$ cannot be found because we are trying to multiply a 2×3 matrix by a 1×2 matrix. The first matrix must have the same number of columns as rows of the second matrix. However, the product $\begin{bmatrix} -2 & 1 \end{bmatrix} \times \begin{bmatrix} 3 & 2 & -1 \\ 0 & 6 & 8 \end{bmatrix}$ can be found.

In general, $AB \neq BA$.

3.1.5.4 Powers of matrices

$A^2 = A \times A$; $A^3 = A \times A \times A$, etc.

If A = $\begin{bmatrix} 3 & 5 \\ -1 & 2 \end{bmatrix}$ then

$\qquad\qquad\qquad\qquad\qquad\qquad\qquad\qquad \begin{bmatrix} 3 & 5 \\ -1 & 2 \end{bmatrix}$

AA

$\qquad\qquad\qquad\qquad\qquad\qquad\qquad\qquad \begin{bmatrix} 4 & 25 \\ -5 & -1 \end{bmatrix}$

A^3

$\qquad\qquad\qquad\qquad\qquad\qquad\qquad\qquad \begin{bmatrix} -13 & 70 \\ -14 & -27 \end{bmatrix}$

3.1.5.5 Operations on CAS

Establish and save the matrices as variables that are available on CAS.

In the examples, the matrices are saved as upper case letters A, B, etc. Some CAS use lower case letters.

i Addition

$\begin{bmatrix} 3 & 5 \\ -1 & 2 \end{bmatrix} \Rightarrow A$

$\qquad\qquad \begin{bmatrix} 3 & 5 \\ -1 & 2 \end{bmatrix}$

$\begin{bmatrix} -1 & 0 \\ 3 & 4 \end{bmatrix} \Rightarrow B$

$\qquad\qquad \begin{bmatrix} -1 & 0 \\ 3 & 4 \end{bmatrix}$

$A+B$

$\qquad\qquad \begin{bmatrix} 2 & 5 \\ 2 & 6 \end{bmatrix}$

$A-B$

$\qquad\qquad \begin{bmatrix} 4 & 5 \\ -4 & -2 \end{bmatrix}$

ii Multiplication by a scalar and calculation of expressions

$2A$

$\qquad\qquad \begin{bmatrix} 6 & 10 \\ -2 & 4 \end{bmatrix}$

$2A+3B$

$\qquad\qquad \begin{bmatrix} 3 & 10 \\ 7 & 16 \end{bmatrix}$

iii Multiplication

AB

$\qquad\qquad \begin{bmatrix} 12 & 20 \\ 7 & 8 \end{bmatrix}$

BA

$\qquad\qquad \begin{bmatrix} -3 & -5 \\ 5 & 23 \end{bmatrix}$

3.1.6 Inverse of a matrix

If A is a **square matrix**, then its **inverse matrix** is denoted by A^{-1}.

A^{-1} **is the matrix such that** $A^{-1}A = AA^{-1} = I$, where I is the identity matrix. This property of inverse matrices makes it possible to solve matrix equations.

For a 2×2 matrix $\begin{bmatrix} a & b \\ c & d \end{bmatrix}$, the inverse is the matrix $\dfrac{1}{ad - bc} \begin{bmatrix} d & -b \\ -c & a \end{bmatrix}$.

The quantity $ad - bc$ is a number called the **determinant**. The determinant of a matrix A can be written as $\det(A)$ or $|A|$. It can also be represented by a delta symbol: Δ.

If $ad - bc = 0$, then $\dfrac{1}{ad - bc}$ will be undefined and **the inverse matrix will not exist**.

Inverses of matrices of order greater than 2×2 are found using CAS.

A matrix whose inverse **can be found** is called an **invertible matrix**.

A matrix whose inverse **does not exist** is called a **singular matrix**.

Using CAS:

The **determinant** of a matrix is an inbuilt function on CAS, det (B), and the inverse of a matrix is found by B^{-1}.

For example, if $A = \begin{bmatrix} 3 & 5 \\ -1 & 2 \end{bmatrix}$ and $B = \begin{bmatrix} -1 & 0 \\ 5 & 4 \end{bmatrix}$, then

$$\det(A) \qquad 11$$

$$\det(B) \qquad -4 \qquad\qquad B\text{^}{-1} \qquad \begin{bmatrix} -1 & 0 \\ 1.25 & 0.25 \end{bmatrix}$$

$$A^{-1} \qquad \begin{bmatrix} \frac{2}{11} & -\frac{5}{11} \\ \frac{1}{11} & \frac{3}{11} \end{bmatrix} \qquad B\text{^}{-1} \qquad \begin{bmatrix} -1 & 0 \\ \frac{5}{4} & \frac{1}{4} \end{bmatrix}$$

3.1.7 Binary matrices

Binary matrices are matrices that have elements that are either '1' or '0'.

Binary matrices can be used to represent a relationship.

Example 6

Consider the numbers 1, 2, 3, 4. A matrix to represent 'is a factor of' would be:

$$\text{Is a factor of} \quad \begin{array}{c} \\ 1 \\ 2 \\ 3 \\ 4 \end{array} \begin{array}{cccc} 1 & 2 & 3 & 4 \\ \begin{bmatrix} 1 & 0 & 0 & 0 \\ 1 & 1 & 0 & 0 \\ 1 & 0 & 1 & 0 \\ 1 & 1 & 0 & 1 \end{bmatrix} \end{array}$$

Other types of matrices that are binary matrices are permutation matrices and dominance matrices.

3.1.7.1 Permutation matrices

Permutation matrices change the order of rows or columns in a matrix. Permutation matrices are always placed before a matrix during multiplication.

A **permutation matrix** is a square matrix where each row or column contains only one element '1' and all the other elements are '0'.

There are only two possible 2 × 2 permutation matrices: $\begin{bmatrix} 1 & 0 \\ 0 & 1 \end{bmatrix}$ and $\begin{bmatrix} 0 & 1 \\ 1 & 0 \end{bmatrix}$.

There are six possible 3 × 3 permutation matrices.

The 3 × 3 identity matrix $\begin{bmatrix} 1 & 0 & 0 \\ 0 & 1 & 0 \\ 0 & 0 & 1 \end{bmatrix}$ and five permutations of the rows of the identity matrix:

$$\begin{bmatrix} 1 & 0 & 0 \\ 0 & 0 & 1 \\ 0 & 1 & 0 \end{bmatrix} \quad \begin{bmatrix} 0 & 1 & 0 \\ 1 & 0 & 0 \\ 0 & 0 & 1 \end{bmatrix} \quad \begin{bmatrix} 0 & 0 & 1 \\ 1 & 0 & 0 \\ 0 & 1 & 0 \end{bmatrix} \quad \begin{bmatrix} 0 & 1 & 0 \\ 0 & 0 & 1 \\ 1 & 0 & 0 \end{bmatrix} \quad \begin{bmatrix} 0 & 0 & 1 \\ 0 & 1 & 0 \\ 1 & 0 & 0 \end{bmatrix}$$

Example 7

Multiply the matrix $A = \begin{bmatrix} 2 & 2 & 2 \\ 3 & 3 & 3 \\ 4 & 4 & 4 \end{bmatrix}$ by the permutation matrix $P = \begin{bmatrix} 0 & 1 & 0 \\ 1 & 0 & 0 \\ 0 & 0 & 1 \end{bmatrix}$.

Solution

$$PA = \begin{bmatrix} 0 & 1 & 0 \\ 1 & 0 & 0 \\ 0 & 0 & 1 \end{bmatrix} \begin{bmatrix} 2 & 2 & 2 \\ 3 & 3 & 3 \\ 4 & 4 & 4 \end{bmatrix} = \begin{bmatrix} 3 & 3 & 3 \\ 2 & 2 & 2 \\ 4 & 4 & 4 \end{bmatrix}$$

The first and second rows of A are exchanged in the product PA. This is because in matrix P, the first and second rows have been exchanged, compared to the identity matrix with 1s along the diagonal.

3.1.7.2 Dominance matrices

If a matrix is constructed to record the results of a round-robin team competition, then the matrix is called a **dominance matrix**. The teams in the competition can be ranked by summing the rows of the matrix, creating a column matrix called a **dominance vector**.

Example 8

Teams W, X, Y and Z competed in a round-robin tournament with the following outcomes:

team W defeated teams X and Y

team X defeated teams Y and Z

team Y defeated team Z

team Z defeated team W.

Rank the teams so that the best team can be established.

Solution

The dominance matrix, D, for the outcome of the 6 matches is shown below where, if team W defeated team X, this is shown as a '1' in the matrix. When Team W lost to team Z this is shown as a '0' in the matrix, etc.

> **Note**
>
> If there are n teams who play each other once, then there are $\frac{n(n-1)}{2}$ matches.

Defeated

$$D = \begin{array}{c} \\ W \\ X \\ Y \\ Z \end{array} \begin{array}{cccc} W & X & Y & Z \\ \begin{bmatrix} 0 & 1 & 1 & 0 \\ 0 & 0 & 1 & 1 \\ 0 & 0 & 0 & 1 \\ 1 & 0 & 0 & 0 \end{bmatrix} \end{array}$$

Summing the rows of the matrix gives the dominance vector (also called the one-step dominance matrix)

$\begin{bmatrix} 2 \\ 2 \\ 1 \\ 1 \end{bmatrix}$, which shows the number of wins for each team. In this case, the dominance vector does not rank

the teams; teams W and X both won two matches.

Second stage wins are, for example, where team W defeated team Y but team Y defeated team Z, so team W is said to have a second-stage win over team Z.

Second stage wins are given in the matrix D^2. This is also called the two-step dominance matrix.

Using CAS to find D^2,

$$D^2 = \begin{bmatrix} 0 & 0 & 1 & 2 \\ 1 & 0 & 0 & 1 \\ 1 & 0 & 0 & 0 \\ 0 & 1 & 1 & 0 \end{bmatrix}$$

The sum of the 1st and 2nd stage wins is $D + D^2 = \begin{bmatrix} 0 & 1 & 2 & 2 \\ 1 & 0 & 1 & 2 \\ 1 & 0 & 0 & 1 \\ 1 & 1 & 1 & 0 \end{bmatrix}$.

The dominance vector is $\begin{bmatrix} 5 \\ 4 \\ 2 \\ 3 \end{bmatrix}$. It gives the total dominance scores of the one-step and two-step

dominance matrices.

The teams are ranked W, X, Z, Y.

3.1.8 Equality of matrices

Two matrices are equal if they have the **same order and the elements in corresponding positions are equal**.

If we are told that two matrices are equal, then we can assume that they have the same order and that elements in the corresponding positions are equal.

For example, if $\begin{bmatrix} a & b \\ c & 2d \end{bmatrix} = \begin{bmatrix} 1 & 2 \\ 3 & 6 \end{bmatrix}$ then $a = 1$, $b = 2$, $c = 3$ and $2d = 6$ … hence $d = 3$.

3.2 Transition matrices

Transition matrices model a system that changes with time.

Example 9

A new sandwich shop, B, has opened in a shopping centre where previously there was only one sandwich shop, A. After a few weeks, a study showed that 75% of those who buy their lunch from Shop A on a given day will buy their lunch from Shop A the next day, whereas 80% of those who buy their lunch from Shop B on a given day will buy their lunch from Shop B the next day. The study also found that Shop A has 70% of the market. What percentage of the lunchtime sandwich market can Shop B expect to acquire in the long term?

This information can be given as a transition diagram.

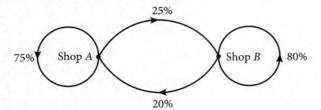

A transition matrix for this situation can be set up.

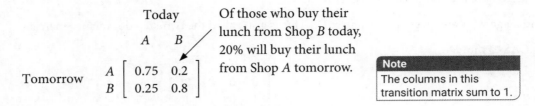

Today

A B

Tomorrow $\begin{array}{c} A \\ B \end{array} \begin{bmatrix} 0.75 & 0.2 \\ 0.25 & 0.8 \end{bmatrix}$

Of those who buy their lunch from Shop B today, 20% will buy their lunch from Shop A tomorrow.

Note
The columns in this transition matrix sum to 1.

The **initial state matrix** for this situation is

$\begin{array}{c} A \\ B \end{array} \begin{bmatrix} 0.7 \\ 0.3 \end{bmatrix}$ ← Initially, Shop A has 70% of the market

Let the transition matrix $T = \begin{bmatrix} 0.75 & 0.2 \\ 0.25 & 0.8 \end{bmatrix}$ and an initial state matrix $S_0 = \begin{bmatrix} 0.7 \\ 0.3 \end{bmatrix}$.

The system that is changing will, after one period of transition (a day in this case), have the state matrix

$$S_1 = TS_0 = \begin{bmatrix} 0.75 & 0.2 \\ 0.25 & 0.8 \end{bmatrix} \begin{bmatrix} 0.7 \\ 0.3 \end{bmatrix} = \begin{bmatrix} 0.585 \\ 0.415 \end{bmatrix}.$$

After two periods of transition (two days) the system will have the state matrix

$$S_2 = TS_1 = \begin{bmatrix} 0.75 & 0.2 \\ 0.25 & 0.8 \end{bmatrix} \begin{bmatrix} 0.585 \\ 0.415 \end{bmatrix} = \begin{bmatrix} 0.5218 \\ 0.4782 \end{bmatrix}$$ and so on.

The **state matrix** after period n can be found using the rule $S_n = T^n S_0$.

From the previous example, $S_2 = TS_1 = T(TS_0) = T^2 S_0 = \begin{bmatrix} 0.75 & 0.2 \\ 0.25 & 0.8 \end{bmatrix}^2 \begin{bmatrix} 0.7 \\ 0.3 \end{bmatrix} = \begin{bmatrix} 0.5218 \\ 0.4782 \end{bmatrix}$

When there is very little change in the values of the state matrices from one state to the next, then we say we have reached **steady state**.

A **steady state** is indicated when $S_{n+1} \approx S_n$ for sufficiently large values of n. **Steady-state proportions** can also be determined by finding T^n for large n.

For the example above, $S_{15} \approx S_{16} = \begin{bmatrix} 0.4444 \\ 0.5556 \end{bmatrix}$, so this is the steady-state matrix.

Alternatively, use CAS to find a high power of T (e.g. 50) to find the steady-state proportions.

Hence, Shop B can expect, in the long term, to acquire 0.5556 of the lunchtime sandwich market. Shop A decreased from 70% to 44%, and Shop B increased from 30% to 56% of the market.

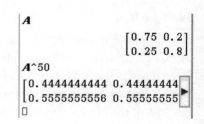

Note
The steady state is independent of the initial state.

3.2.1 Matrix recurrence relations

Matrix recurrence relations can be used to extend modelling to populations that include culling and restocking.

Example 10

A business employs salesmen (S) and technicians (T). Each year these employees can move between the available position or leave the employment (L).

A transition matrix for the employees could be

$$\text{Next Year} \quad \begin{array}{c} \\ S \\ T \\ L \end{array} \begin{array}{ccc} & \text{This year} & \\ S \quad T \quad L \end{array} \left[\begin{array}{ccc} 0.7 & 0.3 & 0 \\ 0.2 & 0.5 & 0 \\ 0.1 & 0.2 & 1 \end{array} \right] = M$$

At the beginning of this year the number of employees is $S_0 = \left[\begin{array}{c} 10 \\ 16 \\ 0 \end{array} \right]$.

$S_1 = MS_0 = \left[\begin{array}{c} 11.8 \\ 10 \\ 4.2 \end{array} \right]$ After one year there are 12 salesmen and 10 technicians. Four employees have left the business.

The total number of employees will decrease over the years so that the manager decides to employ extra staff each year using the matrix $G = \left[\begin{array}{c} 0 \\ 5 \\ 0 \end{array} \right]$ and the number of employees is modelled using $S_{n+1} = MS_n + G$.

$$S_1 = MS_0 + G = \left[\begin{array}{c} 11.8 \\ 15 \\ 4.2 \end{array} \right] \qquad S_2 = MS_1 + G = \left[\begin{array}{c} 12.76 \\ 14.86 \\ 8.38 \end{array} \right] \qquad S_3 = MS_2 + G = \left[\begin{array}{c} 13.39 \\ 14.982 \\ 12.628 \end{array} \right] \text{ and so on.}$$

> **Note**
> The element in row 3 of the S_n matrix represents the total number of staff that have left.

3.2.2 Leslie matrices

A **Leslie matrix** gives a model of population growth, over a period of time, in a closed population. This occurs in an unrestricted environment and usually only the female population is considered as birth rate is one of the components.

The population is divided into age groups and

 i the number in each age group is known

 ii the birth rate for each age group is known

 iii the proportion of each age group that survives the age group (and hence moves into the next age group) is known.

Example 11

A population of female animals has been observed and the number of individuals, their birth rate and survival rate are recorded in the table below.

The animals rarely live beyond 6 years so the animals are divided into three groups, based on ages in time periods of 2 years.

Age group (years)	0–<2	2–<4	4–<6
Number of females	42	28	20
Birthrate	0.32	0.9	0.3
Survival rate	0.65	0.56	0

The Leslie matrix, L, for this information is $L = \begin{bmatrix} 0.32 & 0.9 & 0.3 \\ 0.65 & 0 & 0 \\ 0 & 0.56 & 0 \end{bmatrix}$.

The initial population matrix, S_0, is $S_0 = \begin{bmatrix} 42 \\ 28 \\ 20 \end{bmatrix}$.

The birthrate and survival rate information in the table can be illustrated in the transition diagram below.

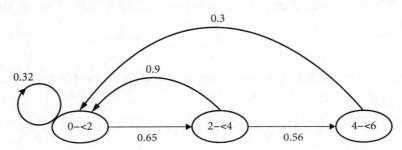

The population after the first time period (2 years) is given by $S_1 = LS_0$:

$$S_1 = \begin{bmatrix} 0.32 & 0.9 & 0.3 \\ 0.65 & 0 & 0 \\ 0 & 0.56 & 0 \end{bmatrix} \begin{bmatrix} 42 \\ 28 \\ 20 \end{bmatrix} = \begin{bmatrix} 45 \\ 27 \\ 16 \end{bmatrix}$$ (Figures are rounded to nearest whole number)

population after 2 time periods (4 years) = $S_2 = LS_1 = \begin{bmatrix} 0.32 & 0.9 & 0.3 \\ 0.65 & 0 & 0 \\ 0 & 0.56 & 0 \end{bmatrix} \begin{bmatrix} 45 \\ 27 \\ 16 \end{bmatrix} = \begin{bmatrix} 44 \\ 29 \\ 15 \end{bmatrix}$

Also, $S_2 = L^2 S_0$ using the general rule:

$$S_{n+1} = LS_n = L^{n+1}S_0$$

Using CAS to find the population after 10 years (5 time periods):

$S_5 = S^5 L_0 = \begin{bmatrix} 45 \\ 27 \\ 16 \end{bmatrix}$; very little change from S_2, so the population could be described as stable.

Glossary

AB ≠ BA In general, AB (does not equal) BA. The particular case of $AB = BA$ is when the number of columns in A equals the number of rows in B. *See* **matrix multiplication**.

A⁻¹ The inverse of matrix A such that $A^{-1}A = AA^{-1} = I$, where I is the identity matrix.

binary matrices Binary matrices have elements that are either '1' or '0'.

column matrix Has order $m \times 1$ with m rows and one column. *See also* **column vector**.

column vector Has order $m \times 1$ with m rows and one column. *See also* **column matrix**.

determinant of a matrix In matrix $\begin{bmatrix} a & b \\ c & d \end{bmatrix}$, $\Delta = ad - bc$.

diagonal matrix A square matrix where all the elements, except those along the leading diagonal, are zero. eg. $\begin{bmatrix} 1 & 0 & 0 \\ 0 & -3 & 0 \\ 0 & 0 & 2 \end{bmatrix}$.

dominance matrices Record the results of a round-robin team competition. The teams in the competition can be ranked by summing the rows of the matrix, creating a column matrix called a dominance vector. *See also* **dominance vector**.

dominance vector Record the results in a round-robin team competition. The teams in the competition can be ranked by summing the rows of the matrix, creating a column matrix called a dominance vector. *See* **dominance matrix**.

elements In a matrix refer to the numbers in the matrix.

equality of matrices Two matrices are equal if they have the **same order** and the **elements in corresponding positions are equal**.

identity matrix Has elements of '1' along the leading diagonal and '0' for the other elements.

- 1×1 identity matrix, I_1, is $\begin{bmatrix} 1 \end{bmatrix}$.

- 2×2 identity matrix, I_2, is $\begin{bmatrix} 1 & 0 \\ 0 & 1 \end{bmatrix}$.

- 3×3 identity matrix, I_3, is $\begin{bmatrix} 1 & 0 & 0 \\ 0 & 1 & 0 \\ 0 & 0 & 1 \end{bmatrix}$.

A+ DIGITAL
Revise this topic's key terms and concepts by scanning the QR code or typing the URL into your browser.

https://get.ga/aplus-vcegeneral-maths

initial state matrix A column matrix with order $m \times 1$.

invertible matrix A matrix whose inverse can be found. This is the case where the determinant of a square matrix is equal to any number except zero.

inverse of a matrix Denoted by A^{-1}.

$A^{-1} = \dfrac{1}{ad - bc} \begin{bmatrix} d & -b \\ -c & a \end{bmatrix}$ for a 2×2 matrix

$\begin{bmatrix} a & b \\ c & d \end{bmatrix}$. The quantity $ad - bc$ is a number called the **determinant**.

Leslie matrix, L Gives a model of population growth, over a period of time, in a closed population. This occurs in an unrestricted environment and usually only the female population is considered as birth rate is one of the components.

matrix A rectangular array of numbers denoted by capital letters A, B …

$A = \begin{bmatrix} 1 & 2 & 3 & 4 \end{bmatrix}$

$B = \begin{bmatrix} 5 & 4 & 7 \\ -2 & 0 & 13 \end{bmatrix}$

matrix operations

- **matrix addition** Only matrices that have the same order can be added or subtracted.

- **matrix division** Does not exist unless in the form of the use of the inverse of matrix to solve a matrix equation.

- **matrix multiplication** The product, AB, of two matrices A and B can only be found if the number of columns in A is equal to the number of rows in B.

- **matrix subtraction** Only matrices that have the same order can be added or subtracted.

matrix recurrence relations Can be used to extend modelling to populations that include culling and restocking.

negative matrices The negative of matrix A is the matrix $-A = -1 \times A$.

order $m \times n$ The matrix has the order $m \times n$ for m rows and n columns.

permutation matrix A square matrix where each row or column contains only one element '1' and all the other elements are '0'. Permutation matrices change the order of rows or columns.

row matrix Matrix has order $1 \times n$ with one row and n columns. See also **row vector**.

row vector Vector has order $1 \times n$ with one row and n columns. See also **row matrix**.

scalar multiplication of matrices If a matrix is multiplied by a scalar (a number) k, then every element in the matrix is multiplied by k.

singular matrix Matrix whose inverse does not exist.

square matrix A matrix with order $m \times m$, or $n \times n$, an equal number of rows and columns.

steady state Indicated when $S_{n+1} \approx S_n$ for sufficiently large values of n.

steady-state proportion Can be determined by finding T^n for large n.

symmetric matrix A square matrix that is equal to its transpose.

For example, $\begin{bmatrix} 3 & 1 \\ 1 & 3 \end{bmatrix}^T = \begin{bmatrix} -1 & 0 \\ 5 & 4 \end{bmatrix}$.

transition matrices Models a system that changes with time. Columns in a transition matrix add up to 1.

transpose of a matrix Denoted by A^T, a transpose of a matrix is where the rows and columns are interchanged; the first row becomes the first column, the second row becomes the second column, etc.

zero matrix Every element of a zero matrix is zero, denoted by 0.

$A \times 0 = 0 \times A = 0$.

Exam practice

Multiple-choice questions

Matrix representation and its applications: 48 questions

Solutions to this section start on page 263.

Question 1

The matrix $\begin{bmatrix} 2 \\ 2 \\ 2 \end{bmatrix}$ could best be described as

A a matrix of order 2 × 3.   **B** a matrix of order 1 × 3.   **C** a row matrix.

D a matrix of order 3 × 2.   **E** a column matrix.

Question 2

Consider the matrix $\begin{bmatrix} 3 & 2 & 3 \\ 3 & 2 & 3 \end{bmatrix}$.

The order of this matrix is

A 2   **B** 3   **C** 2 × 3   **D** 3 × 2   **E** 3 × 3

The following information relates to Questions 3 and 4.

Consider matrix A as shown.

$$\begin{bmatrix} 2 & 3 \end{bmatrix}$$

Question 3

Matrix multiplication $A \times B$ is possible, where B is the matrix

A $\begin{bmatrix} 3 \end{bmatrix}$ **B** $\begin{bmatrix} 3 & 2 \end{bmatrix}$ **C** $\begin{bmatrix} 2 & 3 \end{bmatrix}$

D $\begin{bmatrix} 1 & 2 & 3 \\ 3 & 2 & 1 \end{bmatrix}$ **E** $\begin{bmatrix} 1 & 2 & 2 & 1 \end{bmatrix}$

Question 4

If matrix B is the correct matrix chosen from Question 3, then the order of matrix AB will be

A 1 × 2   **B** 1 × 3   **C** 2 × 1   **D** 3 × 1   **E** 2 × 2

Question 5

Consider matrix P as shown below.

$$\begin{bmatrix} 5 & 3 \end{bmatrix}$$

Matrix addition $P + Q$ is possible, where Q is the matrix

A $\begin{bmatrix} 4 \end{bmatrix}$ **B** $\begin{bmatrix} 3 & 6 & 9 \end{bmatrix}$ **C** $\begin{bmatrix} 1 & 2 & 3 \\ 3 & 2 & 1 \end{bmatrix}$

D $\begin{bmatrix} 1 & 2 & 2 & 1 \end{bmatrix}$ **E** $\begin{bmatrix} 2 & 3 \end{bmatrix}$

Question 6 ○●●●

For the matrix $X = \begin{bmatrix} 7 & 3 & 1 \\ 2 & 4 & 0 \\ -3 & -2 & 5 \end{bmatrix}$, which one of the following is **true** about the element x_{ij}

A $x_{23} = -2$ **B** $x_{13} = -3$ **C** $x_{32} = 0$ **D** $x_{31} = -3$ **E** $x_{21} = 3$

Question 7 ○●●●

The matrix $X = \begin{bmatrix} 1 & -4 & 7 \\ 0 & 1 & 3 \\ 0 & 0 & 1 \end{bmatrix}$.

Which one of the following is **true** about matrix X?

A X is an identity matrix.   **B** $x_{23} = 0$

C X is a triangular matrix.   **D** X is a symmetric matrix.

E X is a diagonal matrix.

Question 8 ○●●●

Consider the matrix $\begin{bmatrix} 1 & 0 \\ 0 & 1 \end{bmatrix}$.

Which one of the following would **not** be true for this matrix?

A Its inverse cannot be found.   **B** It is called an identity matrix.

C It is a symmetric matrix   **D** It is a square matrix.

E It is called a diagonal matrix.

Question 9 ○○●●

If $X = \begin{bmatrix} 1 & 0 \\ a & b \end{bmatrix}$ and $Y = \begin{bmatrix} 2 & 2 \\ 0 & 0 \end{bmatrix}$, then $X - 2Y =$

A $\begin{bmatrix} 3 & 3 \\ a & b \end{bmatrix}$ **B** $\begin{bmatrix} 3 & 2 \\ a & b \end{bmatrix}$ **C** $\begin{bmatrix} -3 & -4 \\ a & b \end{bmatrix}$ **D** $\begin{bmatrix} 2 & 2 \\ 2a & 2a \end{bmatrix}$ **E** $\begin{bmatrix} -1 & -2 \\ a & b \end{bmatrix}$

Question 10 ○○●●

$A = \begin{bmatrix} 1 \\ 2 \\ -1 \end{bmatrix}$, $B = \begin{bmatrix} 5 & -2 & 1 \end{bmatrix}$ and $C = \begin{bmatrix} 2 & -1 & 4 \\ 1 & 0 & -3 \end{bmatrix}$.

Which one of the following is **true** for matrices A, B and C?

A BC will be a 1×3 matrix.   **B** AC will be a 3×2 matrix.   **C** BA will be a 1×1 matrix.

D CA cannot be found.   **E** CB will be a 3×3 matrix.

Question 11 ○○●●

If $A = \begin{bmatrix} 3 & 1 \\ 4 & 6 \end{bmatrix}$ and $B = \begin{bmatrix} 2 & -1 \\ -2 & 0 \end{bmatrix}$, then $2A + 3B$ is equal to

A $\begin{bmatrix} 0 & 5 \\ 2 & 12 \end{bmatrix}$ **B** $\begin{bmatrix} 12 & -1 \\ 2 & 12 \end{bmatrix}$ **C** $\begin{bmatrix} 31 & 0 \\ 12 & 36 \end{bmatrix}$ **D** $\begin{bmatrix} 13 & 1 \\ 8 & 2 \end{bmatrix}$ **E** $\begin{bmatrix} 25 & 0 \\ 10 & 30 \end{bmatrix}$

Question 12 ●●●

A zero matrix can be produced by

A the addition of a matrix and its negative.

B multiplication of a matrix by −1.

C multiplying a matrix by its determinant.

D addition of a row matrix where all elements are '0'.

E addition of a column matrix where all elements are '0'.

Question 13 ●●●

The matrix A^2 will exist if matrix A is

A $\begin{bmatrix} 3 \end{bmatrix}$ 　　　　**B** $\begin{bmatrix} 3 & 2 \end{bmatrix}$ 　　**C** $\begin{bmatrix} 2 & 3 \end{bmatrix}$ 　　**D** $\begin{bmatrix} 1 & 2 & 3 \\ 3 & 2 & 1 \end{bmatrix}$ 　**E** $\begin{bmatrix} 1 & 2 & 2 & 1 \end{bmatrix}$

Question 14 ●●●

If C is the matrix $\begin{bmatrix} 2 & 4 \\ 1 & -1 \end{bmatrix}$, then C^2 is

A $\begin{bmatrix} 2 & 4 \\ 1 & -1 \end{bmatrix}$ 　**B** $\begin{bmatrix} 4 & 16 \\ 1 & 1 \end{bmatrix}$ 　**C** $\begin{bmatrix} 4 & 16 \\ 1 & -1 \end{bmatrix}$ 　**D** $\begin{bmatrix} 8 & 4 \\ 1 & 5 \end{bmatrix}$ 　**E** $\begin{bmatrix} 1 & 5 \\ 8 & 4 \end{bmatrix}$

Question 15 ●●●

For which one of the following matrices is the value of the determinant the largest?

A $\begin{bmatrix} 1 & 2 \\ 3 & 4 \end{bmatrix}$ 　**B** $\begin{bmatrix} 5 & 7 \\ 3 & 0 \end{bmatrix}$ 　**C** $\begin{bmatrix} -1 & -3 \\ 3 & 4 \end{bmatrix}$ 　**D** $\begin{bmatrix} 0 & 2 \\ 3 & 0 \end{bmatrix}$ 　**E** $\begin{bmatrix} 7 & 9 \\ 3 & 4 \end{bmatrix}$

Question 16 ●●●

If A is a 3 × 4 matrix and B is a 2 × 3 matrix, then the number of elements in matrix BA is

A 2 　　　　　**B** 3 　　　　　**C** 4 　　　　　**D** 8 　　　　　**E** 12

Question 17 ●●●

The inverse of the matrix $\begin{bmatrix} w & x \\ y & z \end{bmatrix}$ is

A $\begin{bmatrix} w & x \\ y & z \end{bmatrix}$

B $\begin{bmatrix} \dfrac{w}{wz - xy} & \dfrac{x}{wz - xy} \\ \dfrac{y}{wz - xy} & \dfrac{z}{wz - xy} \end{bmatrix}$

C $\begin{bmatrix} \dfrac{w}{wz - xy} & \dfrac{-x}{wz - xy} \\ \dfrac{-y}{wz - xy} & \dfrac{z}{wz - xy} \end{bmatrix}$

D $\begin{bmatrix} \dfrac{z}{wz - xy} & \dfrac{y}{wz - xy} \\ \dfrac{x}{wz - xy} & \dfrac{w}{wz - xy} \end{bmatrix}$

E $\begin{bmatrix} \dfrac{z}{wz - xy} & \dfrac{-x}{wz - xy} \\ \dfrac{-y}{wz - yx} & \dfrac{w}{wz - xy} \end{bmatrix}$

Question 18 ●●

Which one of the following is **not** true regarding the matrix $A = \begin{bmatrix} 4 & 4 & 5 \\ 2 & 0 & 3 \end{bmatrix}$?

A The matrix A contains 3 columns.

B The product matrix AB exists, where B is any 3×2 matrix.

C Matrix subtraction $A - C$ exists, where C is any 2×3 matrix.

D A^2 does not exist for matrix A.

E The determinant of matrix A is equal to 1.

Question 19 ●●

If $A = \begin{bmatrix} 2 & 3 \\ 4 & 1 \end{bmatrix}$, $B = \begin{bmatrix} 2 \\ 5 \\ 2 \end{bmatrix}$, $C = \begin{bmatrix} 2 \\ -1 \\ 4 \end{bmatrix}$ and $D = \begin{bmatrix} 3 & -1 \\ 3 & -1 \end{bmatrix}$, which one of the following exists?

A $2A - 4B$ **B** $B + C - D$ **C** $3D - A$ **D** $A + B + C + D$ **E** $C - 3A$

Question 20 ●●

If $A = \begin{bmatrix} 3 & 2 \\ 6 & 4 \end{bmatrix}$ and $B = \begin{bmatrix} 6 & 0 \\ 4 & -3 \end{bmatrix}$, which one of the following could **not** be found?

A AB **B** $A + B$ **C** B^2 **D** A^{-1} **E** $-\frac{1}{2}B$

Question 21 ●●

If $I = \begin{bmatrix} 1 & 0 \\ 0 & 1 \end{bmatrix}$, $O = \begin{bmatrix} 0 & 0 \\ 0 & 0 \end{bmatrix}$ and $A = \begin{bmatrix} 1 & 2 \\ 3 & 4 \end{bmatrix}$, which one of the following would **not** be true?

A $I + O = I$ **B** $AO = O$ **C** $A - O = A$ **D** $A + I = A$ **E** $IO = O$

Question 22 ●●

Which one of the following is a singular matrix, a matrix whose inverse cannot be found?

A $\begin{bmatrix} 3 & -3 \\ -1 & 0 \end{bmatrix}$ **B** $\begin{bmatrix} 1 & 0 \\ 1 & 1 \end{bmatrix}$ **C** $\begin{bmatrix} 2 & -2 \\ -2 & -2 \end{bmatrix}$ **D** $\begin{bmatrix} 1 & 4 \\ 6 & 25 \end{bmatrix}$ **E** $\begin{bmatrix} 3 & 3 \\ -3 & -3 \end{bmatrix}$

Question 23 ●●

The inverse of the matrix $\begin{bmatrix} 2 & -1 \\ 1 & -2 \end{bmatrix}$ is

A $\begin{bmatrix} \frac{-2}{3} & \frac{1}{3} \\ \frac{-1}{3} & \frac{2}{3} \end{bmatrix}$ **B** $\begin{bmatrix} -\frac{2}{5} & \frac{1}{5} \\ -\frac{1}{5} & -\frac{2}{5} \end{bmatrix}$ **C** $\begin{bmatrix} 1 & 0 \\ 0 & 1 \end{bmatrix}$ **D** $\begin{bmatrix} \frac{2}{3} & -\frac{1}{3} \\ \frac{1}{3} & -\frac{2}{3} \end{bmatrix}$ **E** $\begin{bmatrix} \frac{2}{5} & -\frac{1}{5} \\ \frac{1}{5} & -\frac{2}{5} \end{bmatrix}$

Question 24 ●●

If $A = \begin{bmatrix} a & b \\ c & d \end{bmatrix}$, $B = \begin{bmatrix} 2 & 3 \\ 0 & 1 \end{bmatrix}$ and $A = 2B$, which one of the following would **not** be true?

A $a = 4$ **B** $b = 6$ **C** $c = 2$ **D** $d = \frac{1}{2}a$ **E** $B = \frac{1}{2}A$

Question 25 ●●

If $\begin{bmatrix} 3 & -2 & 0 \\ 2 & 4 & -1 \end{bmatrix}$ is multiplied by $\begin{bmatrix} 1 & 3 & -1 \\ 2 & 1 & 2 \\ -1 & 1 & 3 \end{bmatrix}$, then the element a_{23} of the product matrix will be

A -3 **B** -1 **C** 3 **D** 7 **E** 9

Question 26 ©VCAA 2014 1MQ6 ●●

The order of matrix X is 3×2.

The element in row i and column j of matrix X is x_{ij} and it is determined by the rule $x_{ij} = i + j$.

The matrix X is

A $\begin{bmatrix} 1 & 2 \\ 3 & 4 \\ 5 & 6 \end{bmatrix}$ **B** $\begin{bmatrix} 2 & 3 \\ 4 & 5 \\ 6 & 7 \end{bmatrix}$ **C** $\begin{bmatrix} 2 & 3 & 4 \\ 3 & 4 & 5 \end{bmatrix}$ **D** $\begin{bmatrix} 1 & 2 \\ 3 & 3 \\ 4 & 4 \end{bmatrix}$ **E** $\begin{bmatrix} 2 & 3 \\ 3 & 4 \\ 4 & 5 \end{bmatrix}$

Question 27 ●●

If $\frac{1}{4}\begin{bmatrix} 4 & 3 \\ 2 & 8 \end{bmatrix} = \frac{1}{2}\begin{bmatrix} a & b \\ 1 & 4 \end{bmatrix}$, then the elements a and b, respectively, will have the values

A $2, \frac{3}{2}$ **B** $2, \frac{2}{3}$ **C** $8, 6$ **D** $\frac{1}{2}, 2$ **E** $1, 4$

Question 28 ●●

If M is the matrix $\begin{bmatrix} a & a & a \\ b & b & b \\ c & c & c \end{bmatrix}$ and XM is the matrix $\begin{bmatrix} c & c & c \\ a & a & a \\ b & b & b \end{bmatrix}$, then X is the permutation matrix

A $\begin{bmatrix} 0 & 0 & 1 \\ 0 & 1 & 0 \\ 1 & 0 & 0 \end{bmatrix}$ **B** $\begin{bmatrix} 0 & 0 & 1 \\ 1 & 0 & 0 \\ 0 & 1 & 0 \end{bmatrix}$ **C** $\begin{bmatrix} 1 & 0 & 0 \\ 0 & 0 & 1 \\ 0 & 1 & 0 \end{bmatrix}$ **D** $\begin{bmatrix} 0 & 0 & 1 \\ 0 & 1 & 1 \\ 1 & 1 & 1 \end{bmatrix}$ **E** $\begin{bmatrix} 0 & 1 & 0 \\ 0 & 0 & 1 \\ 1 & 0 & 0 \end{bmatrix}$

Question 29 ●●

If B is the matrix $\begin{bmatrix} -1 & 0 \\ 1 & 1 \end{bmatrix}$, then $B + B^2$ is equal to

A $\begin{bmatrix} 1 & 0 \\ 0 & 1 \end{bmatrix}$ **B** $\begin{bmatrix} -1 & 0 \\ 0 & 1 \end{bmatrix}$ **C** $\begin{bmatrix} -1 & 0 \\ 1 & 1 \end{bmatrix}$ **D** $\begin{bmatrix} 0 & 0 \\ 1 & 2 \end{bmatrix}$ **E** $\begin{bmatrix} 0 & 0 \\ 1 & 1 \end{bmatrix}$

Question 30 ●●

The inverse of the matrix $\begin{bmatrix} 2 & -3 \\ -1 & x \end{bmatrix}$ is

A $\frac{1}{2x-3}\begin{bmatrix} 2 & -3 \\ -1 & x \end{bmatrix}$ **B** $\frac{1}{2x-3}\begin{bmatrix} x & -3 \\ -1 & 2 \end{bmatrix}$ **C** $\frac{1}{2x-3}\begin{bmatrix} 2 & 3 \\ 1 & x \end{bmatrix}$

D $\frac{1}{2x-3}\begin{bmatrix} x & 3 \\ 1 & 2 \end{bmatrix}$ **E** $\frac{1}{2x+3}\begin{bmatrix} x & 3 \\ 1 & 2 \end{bmatrix}$

Question 31 ●●

If D is the row vector $\begin{bmatrix} 3 & -1 & 2 \end{bmatrix}$ and $I \times D = D$, then I will be the matrix

A $\begin{bmatrix} 1 & 0 & 0 \\ 0 & 1 & 0 \\ 0 & 0 & 1 \end{bmatrix}$
B $\begin{bmatrix} 1 \\ 1 \\ 1 \end{bmatrix}$
C $\begin{bmatrix} 1 & 1 & 1 \\ 1 & 1 & 1 \\ 1 & 1 & 1 \end{bmatrix}$
D $\begin{bmatrix} 1 \end{bmatrix}$
E $\begin{bmatrix} 1 & 0 & 0 \end{bmatrix}$

Question 32 ●●

Which one of the following is **not** true for matrices in general?

A An inverse cannot be found for a matrix whose determinant is zero.

B Inverses can only be found for square matrices.

C Identity matrices are square matrices where the elements on the leading diagonal are one and all the other elements are zero.

D Only matrices with the same order can be multiplied.

E Addition and subtraction can only be carried out with matrices of the same order.

Question 33 ●●

Consider the matrix $\begin{bmatrix} a & 8 \\ 5 & 4 \end{bmatrix}$. The inverse of this matrix will not exist if a equals

A 2 **B** 4 **C** 6 **D** 8 **E** 10

Question 34 ●●

Which one of the following is the inverse of the matrix $\begin{bmatrix} 1 & 2 & -1 \\ 3 & 0 & -2 \\ 1 & 1 & -1 \end{bmatrix}$?

A $\begin{bmatrix} 2 & 1 & -4 \\ 1 & 0 & -1 \\ 3 & 1 & -6 \end{bmatrix}$
B $\begin{bmatrix} -1 & 2 & 1 \\ 3 & 0 & -2 \\ -1 & 1 & 1 \end{bmatrix}$
C $\begin{bmatrix} 1 & 0 & 0 \\ 0 & 1 & 0 \\ 0 & 0 & 1 \end{bmatrix}$

D $\begin{bmatrix} -1 & -2 & 1 \\ -3 & 0 & 2 \\ -1 & -1 & 1 \end{bmatrix}$
E The inverse cannot be found.

Question 35 ©VCAA 2009 1MQ5 ●●

A, *B*, *C*, *D* and *E* are five intersections joined by roads as shown in the diagram below.

Some of these roads are one-way only.

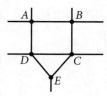

The matrix below indicates the direction that cars travel along each of these roads.

In this matrix

- '1' in column *A* and row *B* indicates that cars can travel directly from *A* to *B*

- '0' in column *B* and row *A* indicates that cars cannot travel directly from *B* to *A* (either it is a one-way road or no road exists).

From intersection

$$\begin{array}{c} \begin{array}{ccccc} A & B & C & D & E \end{array} \\ \begin{bmatrix} 0 & 0 & 0 & 0 & 0 \\ 1 & 0 & 0 & 0 & 0 \\ 0 & 1 & 0 & 1 & 1 \\ 1 & 0 & 0 & 0 & 0 \\ 0 & 0 & 1 & 1 & 0 \end{bmatrix} \begin{array}{l} A \\ B \\ C \\ D \\ E \end{array} \end{array}$$ To intersection

Cars can travel in both directions between intersections

A *A* and *D*   **B** *B* and *C*   **C** *C* and *D*   **D** *D* and *E*   **E** *C* and *E*

Question 36 ●●

The results of a round-robin competition between four teams are given below:

team *A* defeated team *B*

team *B* defeated team *C*

team *C* defeated teams *A* and *D*

team *D* defeated teams *A* and *B*.

Which one of the following is the dominance matrix for the results of the competition?

A $\begin{bmatrix} 0 & 1 & 0 & 0 \\ 0 & 0 & 1 & 0 \\ 1 & 0 & 0 & 1 \\ 1 & 0 & 0 & 0 \end{bmatrix}$ **B** $\begin{bmatrix} 0 & 1 & 0 & 0 \\ 0 & 0 & 1 & 0 \\ 1 & 0 & 0 & 1 \\ 1 & 1 & 0 & 0 \end{bmatrix}$ **C** $\begin{bmatrix} 0 & 1 & 0 & 0 \\ 0 & 0 & 1 & 0 \\ 1 & 1 & 0 & 1 \\ 1 & 1 & 0 & 0 \end{bmatrix}$

D $\begin{bmatrix} 0 & 1 & 1 & 1 \\ 1 & 0 & 1 & 0 \\ 1 & 1 & 0 & 1 \\ 1 & 0 & 1 & 0 \end{bmatrix}$ **E** $\begin{bmatrix} 0 & 1 & 0 & 1 \\ 0 & 0 & 1 & 1 \\ 1 & 1 & 0 & 1 \\ 1 & 1 & 1 & 1 \end{bmatrix}$

The following information relates to Questions 37 and 38.

The results of a round-robin competition between five teams is summarised in the dominance matrix below, where team A defeating team B is represented by a '1'.

$$
\begin{array}{c}
 \\
A \\
B \\
C \\
D \\
E
\end{array}
\begin{array}{c}
\begin{array}{ccccc} A & B & C & D & E \end{array} \\
\left[
\begin{array}{ccccc}
0 & 1 & 1 & 0 & 0 \\
0 & 0 & 0 & 1 & 1 \\
0 & 1 & 0 & 0 & 1 \\
1 & 0 & 1 & 0 & 1 \\
1 & 0 & 0 & 0 & 0
\end{array}
\right]
\end{array}
$$

Question 37 ●●

Using the matrix above, it can be shown that team C will have two-step dominance over

A teams A, B, and E.   **B** teams A, D and E.   **C** teams A, B, and D.

D teams B, D and E.   **E** teams B and E.

Question 38 ●●

Using the total dominance scores, the ranking of the teams is

A D first, B second, C third, A fourth and E last.

B D first, C second, A third, B fourth and E last.

C D first, A second, C third, B fourth and E last.

D D first, A and B equal second, C fourth and E last.

E D first, A and C equal second, B fourth and E last.

Question 39 ©VCAA 2014 1MQ8 ●●●

Wendy will have lunch with one of her friends each day of this week.

Her friends are Angela (A), Betty (B), Craig (C), Daniel (D) and Edgar (E).

On Monday, Wendy will have lunch with Craig.

Wendy will use the transition matrix below to choose a friend to have lunch with for the next four days of the week.

$$
\begin{array}{c}
 \\
 \\
T = \\
 \\

\end{array}
\begin{array}{c}
\text{Today} \\
\begin{array}{ccccc} A & B & C & D & E \end{array} \\
\left[
\begin{array}{ccccc}
0 & 1 & 0 & 0 & 0 \\
0 & 0 & 0 & 1 & 0 \\
1 & 0 & 0 & 0 & 0 \\
0 & 0 & 0 & 0 & 1 \\
0 & 0 & 1 & 0 & 0
\end{array}
\right]
\end{array}
\begin{array}{l}
A \\
B \\
C \quad \text{Tomorrow} \\
D \\
E
\end{array}
$$

The order in which Wendy has lunch with her friends for the next four days is

A Angela, Betty, Craig, Daniel   **B** Daniel, Betty, Angela, Craig

C Daniel, Betty, Angela, Edgar   **D** Edgar, Angela, Daniel, Betty

E Edgar, Daniel, Betty, Angela

The following information relates to Questions 40 and 41.

The results of a round-robin competition between four teams are represented by the network diagram shown.

On the diagram, $\overset{A}{\bullet}\longrightarrow\overset{B}{\bullet}$ means that team A defeated team B.

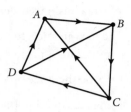

Question 40 ○○●

Which one of the following is the adjacency matrix for the competition?

A $\begin{bmatrix} 0 & 1 & 0 & 0 \\ 0 & 0 & 1 & 0 \\ 1 & 0 & 0 & 1 \\ 1 & 0 & 0 & 0 \end{bmatrix}$

B $\begin{bmatrix} 0 & 1 & 0 & 0 \\ 0 & 0 & 1 & 0 \\ 1 & 0 & 0 & 1 \\ 1 & 1 & 0 & 0 \end{bmatrix}$

C $\begin{bmatrix} 0 & 1 & 0 & 0 \\ 0 & 0 & 1 & 0 \\ 1 & 1 & 0 & 1 \\ 1 & 1 & 0 & 0 \end{bmatrix}$

D $\begin{bmatrix} 0 & 1 & 1 & 1 \\ 1 & 0 & 1 & 0 \\ 1 & 1 & 0 & 1 \\ 1 & 0 & 1 & 0 \end{bmatrix}$

E $\begin{bmatrix} 0 & 1 & 0 & 1 \\ 0 & 0 & 1 & 1 \\ 1 & 1 & 0 & 1 \\ 1 & 1 & 1 & 1 \end{bmatrix}$

Question 41 ○○●

The final ranking of the teams, first to last, will be

A C, D, B, A **B** A, C, D, B **C** D, C, A, B

D C, D, A, B **E** cannot be determined

Question 42 ○○○

P and Q are square matrices whose inverses exist.

Which one of the following is **not** true for P and Q if $P \neq Q$?

A $P^3 = P \times P \times P$ **B** $PQP = P^2Q$ **C** $QQ^{-1} = Q^{-1}Q$

D $P^2P^{-1} = P$ **E** $Q^{-1}Q^3 = Q^2$

Question 43 ○○○

A is a 2×2 matrix, whose inverse exists. I is the 2×2 identity matrix.

Which one of the following is **not** correct?

A $AA^{-1} = I$ **B** $(A - I)I = A - I$ **C** $IAA^{-1}I^{-1} = I$ **D** $(A^{-1})^{-1} = A$ **E** $(AA^{-1})^{-1} = A^{-1}$

Question 44 ○○○

If the matrix B^2 is $\begin{bmatrix} 4 & 4 \\ 0 & 4 \end{bmatrix}$, then the original matrix B could be

A $\begin{bmatrix} 2 & 2 \\ 0 & 2 \end{bmatrix}$ **B** $\begin{bmatrix} 2 & 1 \\ 0 & 2 \end{bmatrix}$ **C** $\begin{bmatrix} 2 & 1 \\ 1 & 2 \end{bmatrix}$ **D** $\begin{bmatrix} 2 & 0 \\ 0 & 2 \end{bmatrix}$ **E** $\begin{bmatrix} 1 & 2 \\ 2 & 1 \end{bmatrix}$

Question 45 ●●●

When a business ordered 7 boxes of paper and 8 printer cartridges, they were invoiced for a total of $816.50.

The next invoice, for a total of $293.50, was for an order of 10 boxes of paper but this invoice included a refund for 3 printer cartridges that were returned.

Let x be the price of a box of paper and y be the price of a printer cartridge.

A matrix equation that could be used to calculate the values of x and y would be

A $\begin{bmatrix} 7 & 8 \\ 10 & 3 \end{bmatrix} \begin{bmatrix} x \\ y \end{bmatrix} = \begin{bmatrix} 816.5 \\ 293.5 \end{bmatrix}$

B $\begin{bmatrix} x \\ y \end{bmatrix} \begin{bmatrix} 7 & 8 \\ 10 & 3 \end{bmatrix} = \begin{bmatrix} 816.5 \\ 293.5 \end{bmatrix}$

C $\begin{bmatrix} x & y \end{bmatrix} \begin{bmatrix} 7 & 8 \\ 10 & 3 \end{bmatrix} = \begin{bmatrix} 816.5 \\ 293.5 \end{bmatrix}$

D $\begin{bmatrix} x \\ y \end{bmatrix} \begin{bmatrix} 7 & 8 \\ 10 & -3 \end{bmatrix} = \begin{bmatrix} 816.5 \\ 293.5 \end{bmatrix}$

E $\begin{bmatrix} 7 & 8 \\ 10 & -3 \end{bmatrix} \begin{bmatrix} x \\ y \end{bmatrix} = \begin{bmatrix} 816.5 \\ 293.5 \end{bmatrix}$

Question 46 ●●●

If $\begin{bmatrix} -3 \\ 4 \end{bmatrix} \begin{bmatrix} a & b \end{bmatrix} = \begin{bmatrix} 6 & x \\ y & 12 \end{bmatrix}$, then the values of x and y, respectively are

A $2, 3$ **B** $-9, 8$ **C** $-9, -8$ **D** $9, -8$ **E** $-2, 3$

Question 47 ●●●

If $\begin{bmatrix} 5 & 2 \\ 8 & 4 \end{bmatrix} B = \begin{bmatrix} 2 & 0 \\ 0 & 2 \end{bmatrix}$, then $B =$

A $\dfrac{1}{8}\begin{bmatrix} 4 & -2 \\ -8 & 5 \end{bmatrix}$ **B** $\dfrac{1}{4}\begin{bmatrix} 4 & -2 \\ -8 & 5 \end{bmatrix}$ **C** $\dfrac{1}{2}\begin{bmatrix} 4 & -2 \\ -8 & 5 \end{bmatrix}$ **D** $2\begin{bmatrix} 4 & -2 \\ -8 & 5 \end{bmatrix}$ **E** $\begin{bmatrix} 4 & -2 \\ -8 & 5 \end{bmatrix}$

Question 48 ●●●

For matrices A and B, the product $AB = \begin{bmatrix} 9 & 0 & 0 \\ 0 & 9 & 0 \\ 0 & 0 & 9 \end{bmatrix}$. If B is the matrix $\begin{bmatrix} 7 & 2 & 6 \\ 9 & 0 & 9 \\ -1 & 1 & 3 \end{bmatrix}$ then the inverse of matrix A will be

A $\dfrac{1}{9}B$ **B** $\dfrac{1}{3}B$ **C** $3B$

D $9B$ **E** cannot be determined

Transition matrices: 17 questions

Solutions to this section start on page 267.

Question 49 ●●○

The Tri-Hi Supermarket has 34% of the market share but has found that only 70% of the customers who shop at Tri-Hi in one week will return to shop at Tri-Hi the following week. Of the customers who shop at another supermarket in any week, 10% will shop at Tri-Hi in the following week. A transition matrix for this situation could be

A $\begin{bmatrix} 0.34 & 0.7 \\ 0.66 & 0.3 \end{bmatrix}$ **B** $\begin{bmatrix} 0.34 & 0.9 \\ 0.66 & 0.1 \end{bmatrix}$ **C** $\begin{bmatrix} 0.7 & 0.1 \\ 0.3 & 0.9 \end{bmatrix}$

D $\begin{bmatrix} 0.7 & 0.9 \\ 0.3 & 0.1 \end{bmatrix}$ **E** $\begin{bmatrix} 0.34 & 0.1 \\ 0.66 & 0.9 \end{bmatrix}$

Question 50 ●●○

A system under transition has the transition matrix:

$$\text{Next} \begin{array}{c} \\ A \\ B \end{array} \begin{array}{c} \overset{\displaystyle \text{Now}}{\overset{A \quad B}{}} \\ \begin{bmatrix} 0.7 & 0.2 \\ 0.3 & 0.8 \end{bmatrix} \end{array}$$

The steady-state proportion for A will be

A 0.2   **B** 0.3   **C** 0.4   **D** 0.6   **E** 0.7

Question 51 ●●○

A system undergoing change has a transition matrix of

$$\text{Next} \begin{array}{c} X \\ Y \end{array} \overset{\overset{\displaystyle \text{Now}}{X \quad Y}}{\begin{bmatrix} 0.68 & 0.4 \\ 0.32 & 0.6 \end{bmatrix}} \text{ and its initial state matrix is } \begin{array}{c} X \\ Y \end{array} \begin{bmatrix} 168 \\ 232 \end{bmatrix}.$$

The state of the system after three transition periods will be

A $\begin{bmatrix} 207 \\ 193 \end{bmatrix}$ **B** $\begin{bmatrix} 218 \\ 182 \end{bmatrix}$ **C** $\begin{bmatrix} 221 \\ 179 \end{bmatrix}$ **D** $\begin{bmatrix} 222 \\ 178 \end{bmatrix}$ **E** $\begin{bmatrix} 225 \\ 175 \end{bmatrix}$

The following information relates to Questions 52, 53 and 54.

Car rental company A currently holds 22% of the car rental market. Research has shown that 60% of the customers who rent a car from A in one month will rent a car from A in the following month and 10% of those who rented a car from another company in the previous month will rent from A in the following month.

Question 52 ●●●

The transition matrix and initial state matrix, respectively, for this changing situation could be

A $\text{Next} \begin{array}{c} A \\ \text{Other} \end{array} \overset{\overset{\displaystyle \text{Now}}{A \quad \text{Other}}}{\begin{bmatrix} 0.6 & 0.4 \\ 0.1 & 0.9 \end{bmatrix}}$ and $\begin{array}{c} A \\ \text{Other} \end{array} \begin{bmatrix} 0.22 \\ 0.78 \end{bmatrix}$ **B** $\text{Next} \begin{array}{c} A \\ \text{Other} \end{array} \overset{\overset{\displaystyle \text{Now}}{A \quad \text{Other}}}{\begin{bmatrix} 0.6 & 0.1 \\ 0.4 & 0.9 \end{bmatrix}}$ and $\begin{array}{c} A \\ \text{Other} \end{array} \begin{bmatrix} 0.22 \\ 0.78 \end{bmatrix}$

C $\text{Next} \begin{array}{c} A \\ \text{Other} \end{array} \overset{\overset{\displaystyle \text{Now}}{A \quad \text{Other}}}{\begin{bmatrix} 0.6 & 0.1 \\ 0.4 & 0.9 \end{bmatrix}}$ and $\begin{array}{c} A \\ \text{Other} \end{array} \begin{bmatrix} 1 \\ 0 \end{bmatrix}$ **D** $\text{Next} \begin{array}{c} A \\ \text{Other} \end{array} \overset{\overset{\displaystyle \text{Now}}{A \quad \text{Other}}}{\begin{bmatrix} 0.9 & 0.4 \\ 0.1 & 0.6 \end{bmatrix}}$ and $\begin{array}{c} A \\ \text{Other} \end{array} \begin{bmatrix} 0.22 \\ 0.78 \end{bmatrix}$

E $\text{Next} \begin{array}{c} A \\ \text{Other} \end{array} \overset{\overset{\displaystyle \text{Now}}{A \quad \text{Other}}}{\begin{bmatrix} 0.6 & 0.1 \\ 0.4 & 0.9 \end{bmatrix}}$ and $\begin{array}{c} A \\ \text{Other} \end{array} \begin{bmatrix} 0.22 \\ 0.78 \end{bmatrix}$

Question 53 ●●●

Assuming that the choice of car rental company is entirely dependent on the choice in the previous month, Apex's percentage share of the car rental market after two months will be

A 20%   **B** 20.5%   **C** 21%   **D** 22%   **E** 24%

Question 54 ●●●

If this situation continues, then the steady-state percentage share of the car rental market for the Apex company is expected to be

A 20%   **B** 21%   **C** 22%   **D** 23%   **E** 24%

The following information relates to Questions 55, 56 and 57.

A large population of mutton birds migrates each year to a remote island to nest and breed. There are four nesting sites on the island, A, B, C and D.

Researchers suggest that the following transition matrix can be used to predict the number of mutton birds nesting at each of the four sites in subsequent years. An equivalent transition diagram is also given.

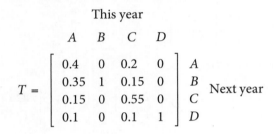

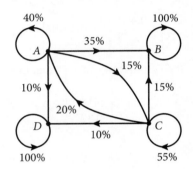

Question 55 ©VCAA 2008 1MQ7 ●○○

Two thousand eight hundred mutton birds nest at site C in 2008.

Of these 2800 birds, the number that nest at site A in 2009 is predicted to be

A 560   **B** 980   **C** 1680   **D** 2800   **E** 3360

Question 56 ©VCAA 2008 1MQ8 ●●○

The transition matrix predicts that, in the long term, the mutton birds will

A nest only at site A.   **B** nest only at site B.

C nest only at sites A and C.   **D** nest only at sites B and D.

E continue to nest at all four sites.

Question 57 ©VCAA 2008 1MQ9 ●●●

Six thousand mutton birds nest at site B in 2008.

Assume that an equal number of mutton birds nested at each of the four sites in 2007. The same transition matrix applies.

The total number of mutton birds that nested on the island in 2007 was

A 6000   **B** 8000   **C** 12 000   **D** 16 000   **E** 24 000

The following information relates to Questions 58, 59 and 60.

In a particular region, in Spring, the weather is changeable from day-to-day.

The following transition matrix has been constructed from records to model the changing conditions.

$$
\begin{array}{cc}
 & \text{Today} \\
 & \text{Fine\quad Showers}
\end{array}
$$

$$
\text{Next day}\quad
\begin{array}{c}
\text{Fine} \\
\text{Showers}
\end{array}
\begin{bmatrix}
0.35 & 0.4 \\
0.65 & 0.6
\end{bmatrix}
$$

Question 58 ⬤⬤⬤

An interpretation of the element 0.65 in the transition matrix is

A 65% of the days will be fine.

B 65% of the days that are fine today will have showers tomorrow.

C 65% of the days that have showers today will be fine tomorrow.

D 65% of the days that have showers today will have showers tomorrow.

E 65% of all days will have showers.

Question 59 ⬤⬤⬤

If today is fine, then the initial state matrix is $S_0 = \begin{bmatrix} 1 \\ 0 \end{bmatrix}$. Assuming that the weather on any day is entirely

dependent on the weather of the previous day, the chance of it being fine in two days' time, correct to two decimal places, will be

A 0.32   **B** 0.35   **C** 0.38   **D** 0.40   **E** 0.65

Question 60 ⬤⬤⬤

The steady-state proportions of this situation for fine days and days with showers, respectively, are

A $\dfrac{1}{2}, \dfrac{1}{2}$ **B** $\dfrac{8}{21}, \dfrac{13}{21}$ **C** $\dfrac{13}{21}, \dfrac{8}{21}$ **D** $\dfrac{7}{20}, \dfrac{13}{20}$ **E** $\dfrac{13}{20}, \dfrac{7}{20}$

Question 61 ©VCAA 2009 1MQ9 ⬤⬤⬤

$T = \begin{bmatrix} 0.8 & 0.3 \\ 0.2 & 0.7 \end{bmatrix}$ is a transition matrix.

$S_3 = \begin{bmatrix} 1150 \\ 850 \end{bmatrix}$ is a state matrix.

If $S_3 = TS_2$, then S_2 equals

A $\begin{bmatrix} 1000 \\ 1000 \end{bmatrix}$ **B** $\begin{bmatrix} 1090 \\ 940 \end{bmatrix}$ **C** $\begin{bmatrix} 1100 \\ 900 \end{bmatrix}$ **D** $\begin{bmatrix} 1150 \\ 850 \end{bmatrix}$ **E** $\begin{bmatrix} 1175 \\ 825 \end{bmatrix}$

Leslie matrices: 4 questions

Solutions to this section start on page 269.

Question 62

The following information is to be used to produce a Leslie matrix for a closed population of female mammals who do not live beyond 3 years.

Age (years)	0 < 1	1–<2	2–3
Number of females	43	56	32
Birthrate	1.3	3.2	0.8
Survival rate	0.54	0.68	0

The Leslie matrix for this information would be

A $\begin{bmatrix} 43 & 56 & 32 \\ 1.3 & 3.2 & 0.8 \\ 0.54 & 0.68 & 0 \end{bmatrix}$
B $\begin{bmatrix} 1.3 & 3.2 & 0.8 \\ 0.54 & 0.68 & 0 \end{bmatrix}$
C $\begin{bmatrix} 43 & 1.3 & 0.54 \\ 56 & 3.2 & 0.68 \\ 32 & 0.8 & 0 \end{bmatrix}$

D $\begin{bmatrix} 1.3 & 3.2 & 0.8 \\ 0.54 & 0 & 0 \\ 0 & 0.68 & 0 \end{bmatrix}$
E $\begin{bmatrix} 1.3 & 0.54 & 0 \\ 3.2 & 0 & 0.68 \\ 0.8 & 0 & 0 \end{bmatrix}$

The following information refers to Questions 63, 64 and 65.

A closed female population has been observed over the 9 years of their life span. The population has been grouped by age into three groups and the survival and reproduction rates of each group has been observed.

The table below gives the age groups and the initial population.

Age group	0–<3	3–<6	6–9
Initial population	43	56	32

The Leslie matrix for this female population is $\begin{bmatrix} 0.36 & 0.8 & 0.25 \\ 0.47 & 0 & 0 \\ 0 & 0.52 & 0 \end{bmatrix}$.

Question 63

The element 0.47 in the Leslie matrix given can be interpreted as

A 47% of the 0–<3 age group will reproduce in the next 3 years.

B 47% of the 0–<3 age group will die in the next 3 years.

C 47% of the 0–<3 age group will survive in the next 3 years.

D 53% of the 0–<3 age group will reproduce in the next 3 years.

E 53% of the 0–<3 age group will move to the next age group in the next 3 years.

Question 64

The population of individuals, to the nearest individual, in the 0–<3 group after 3 years is

A 68   **B** 20   **C** 49   **D** 46   **E** 48

Question 65

The population of females in the 3–<6 age group and the 6–<9 age group after 12 years compared to the initial populations in these age groups, respectively, will

A both stay the same.   **B** increase and decrease.   **C** decrease and increase.

D increase and increase.   **E** decrease and decrease.

Extended-answer questions

Solutions to this section start on page 269.

66 Car hire (15 marks)

Question 66.1 (6 marks)

Let A be the matrix $\begin{bmatrix} 3 & 2 & 1 \\ 4 & 3 & 6 \\ 5 & 2 & 3 \end{bmatrix}$.

a   **i** Multiply A by the matrix $\begin{bmatrix} 1 \\ 1 \\ 1 \end{bmatrix}$.   2 marks

  ii Explain the effect on the elements of A by this product of matrices.   1 mark

b   **i** Find the matrix AB if $B = \begin{bmatrix} 1 & 0 & 0 \\ 0 & 2 & 0 \\ 0 & 0 & 3 \end{bmatrix}$.   2 marks

  ii Explain the effect on the elements of A by this product of matrices.   1 mark

Question 66.2 (4 marks)

A city car-rental company hires out cars for 1, 2, 3, 4 or 5 days.

The following table gives the cost per day, in dollars, of hiring a small, medium, large or four-wheel-drive car.

Car	Cost per day (Dollars)				
	1 day	2 days	3 days	4 days	5 days
Small	50	45	40	37	35
Medium	60	55	50	47	45
Large	70	65	60	57	55
4WD	80	75	70	67	65

a Write down the values in this table as a 4 × 5 cost per day matrix, C.   1 mark

b Write down a matrix T such that the product CT will give a 4 × 5 matrix whose elements represent the total cost of hiring each of the types of cars for 1, 2, 3, 4 or 5 days.   2 marks

c Calculate and write down the product CT.   1 mark

Question 66.3 (5 marks)

The following data about small car hire was collected over three months.

Hiring period	Month 1	Month 2	Month 3
1 day	58	47	62
2 days	114	121	141
3 days	127	108	127
4 days	59	61	73
5 days	43	23	38

a Write a 5 × 3 matrix, H, that represents the number of cars that were hired for each of the 1, 2, 3, 4, or 5 days for these months.   1 mark

b Write down a suitable matrix that when multiplied by H will give a matrix whose elements represent the total number of hirings of small cars for each of the 1, 2, 3, 4 or 5 days over the three-month period. Multiply H by this matrix.   2 marks

c Use your matrix from Question **66.2** part **c** to write a row matrix that represents the cost of hiring a small car for each of the 1, 2, 3, 4 or 5 days.   1 mark

d Use your matrices from parts **b** and **c** to calculate the total revenue received from the hiring of small cars over these three months.   1 mark

67 Changing cars (15 marks)

Question 67.1 (2 marks)

Let A be the matrix $\begin{bmatrix} 3 & 2 & 1 \\ 4 & 3 & 6 \\ 5 & 2 & 3 \end{bmatrix}$.

a Multiply $\begin{bmatrix} 1 & 1 & 1 \end{bmatrix}$ by A.   1 mark

b Explain the effect on the elements of A by this product of matrices.   1 mark

Question 67.2 (3 marks)

A city car-rental company hires out cars for 1, 2, 3, 4 or 5 days.

The following table gives the number of rentals for each type of car for 1, 2, 3, 4 or 5 days over a month.

	Number of rentals			
	Small	Medium	Large	4WD
1 day	58	60	35	11
2 days	114	108	132	27
3 days	127	148	96	52
4 days	59	96	112	48
5 days	43	29	89	53

a Write down the values in this table as a 5 × 4 number of rentals matrix, N.   1 mark

b Write down a matrix S such that the product SN will give a 1 × 4 matrix whose elements represent the total number of rentals in the month for each of Small, Medium, Large or 4WD vehicles.   1 mark

c Calculate the matrix SN.   1 mark

Question 67.3 (10 marks)

A-A car rental company has only one other competitor, B-B, in their city. At present, A-A has 55% of the rental market and B-B has 45%.

A-A has started an advertising campaign in order to increase its market share.

After several weeks of this campaign, some data recorded showed that 90% of customers who rented from A-A this time rented from A-A last time and 40% of those who rented from B-B last time rented from A-A this time.

a Use the information above to complete the following table showing the change from
   one week to the next.   2 marks

		Last rental	
		A-A	**B-B**
This rental	**A-A**	0.9	
	B-B		

b From the table above, write a transition matrix, T.   1 mark

c If A-A started with 55% of the market, write down an initial state column matrix S_0.   1 mark

d Assuming that the choice of rental company depended entirely on the choice in
   the previous rental,

   i find the percentage, correct to one decimal place, of the market captured by
   A-A after one transition period   2 marks

   ii find the percentage, correct to one decimal place, of the market captured by
   A-A after five transition periods.   2 marks

e After how many transition periods will T reach a steady state?   1 mark

f What is the steady-state percentage for the company A-A?   1 mark

68 Fuel for thought (12 marks)

Question 68.1 (7 marks)

In recent times, fuel prices have changed from day to day. At one particular fuel station that sells unleaded petrol (U), liquefied petroleum gas (LPG) and diesel fuel (D), the quantity sold, in litres, of each of these fuels is recorded for three days along with the total takings for these three days. The results are given in the table below.

	Fuel Type			
Day	**U**	**LPG**	**D**	**Takings ($)**
1	16 800	10 890	5660	32 839.55
2	15 400	12 630	8320	35 137.85
3	14 250	9 980	7160	31 103.45

a Write a 3 × 3 matrix, Q, that represents the quantity of each type of fuel sold on each
   of the three days.   1 mark

The proprietor of the fuel shop cannot remember the price of the fuels on each of the three days, so he wishes to get a representative price for the three days.

b If u, g and d are the representative prices for the three days of a litre of unleaded, LPG and diesel fuel respectively, write a 3×1 matrix, X, that represents these prices over the three days. — 1 mark

c Write a 3×1 matrix, T, that represents the takings over the three days. — 1 mark

d Set up a matrix equation that will enable you to find the values of u, g and d. — 2 marks

Question 68.2 (7 marks)

A light rail connection to the city has recently been built from an outlying suburb where previously most of the commuters travelled by private car.

After several weeks, a survey was taken of regular commuters and the following data was recorded.

Of those who travelled on the light rail last week, 80% travelled on the light rail this week, 10% travelled by private car and 10% took another form of transport.

Of those who travelled by private car last week, 70% travelled by private car this week, 10% travelled on the light rail and 20% took another form of transport.

Of those who took another form of transport last week, 30% took another form of transport this week, 50% travelled by light rail and 20% travelled by private car.

a Using the information above, complete the following table. — 2 marks

		Last week (%)		
		Light rail	**Private car**	**Other**
This week (%)	**Light rail**	80		50
	Private car	10	70	
	Other			

b Write a transition matrix, T, from the figures in the table. — 1 mark

c If, at the time of the survey, 40% of commuters were travelling on the light rail, 50% were using a private car and 10% were using another form of transport to commute to the city, write an initial state column matrix, S_0, for this situation. — 1 mark

d Assuming that the choice of transport in one week depended entirely on the choice of transport in the previous week, find the percentage of commuters that will be using the light rail after

  i one week — 1 mark

  ii five weeks. — 1 mark

  Give your answers correct to one decimal place.

e Find the steady-state percentage of commuters that will be using the light rail. Give your answer correct to the nearest percentage. — 1 mark

69 Matrix manoeuvres (12 marks)

Question 69 (12 marks)

If X is the matrix $\begin{bmatrix} 3 & 0 & 2 \\ 1 & -1 & 4 \\ -2 & 1 & 0 \end{bmatrix}$, find

a the matrix $2X$ 1 mark

b the matrix X^2 1 mark

c X^T, the transpose of matrix X 1 mark

d the product of X and the matrix $\begin{bmatrix} \frac{1}{2} & 0 & 0 \\ 0 & \frac{1}{2} & 0 \\ 0 & 0 & \frac{1}{2} \end{bmatrix}$ 2 marks

e the matrix A such that $2X - A = \begin{bmatrix} 5 & 4 & -1 \\ 2 & 1 & 3 \\ -3 & 4 & -5 \end{bmatrix}$ 2 marks

f the inverse matrix, X^{-1}, of X

   i expressing the elements in this inverse matrix as fractions   1 mark

   ii writing your answer as kB, where k is a constant and D is a matrix whose elements
   are whole numbers   1 mark

g the matrix D such that $DX = \begin{bmatrix} 14 & 0 & 0 \\ 0 & 14 & 0 \\ 0 & 0 & 14 \end{bmatrix}$.   3 marks

70 Mixes (12 marks)

Question 70.1 (6 marks)

Laura's Lollies makes three mixes of popular sweets.

   The Fruit mix contains 50% fruit jellies, 25% chocolates and 25% chews.

   The Chocolate mix contains 30% fruit jellies, 45% chocolates and 25% chews.

   The Lasting mix contains 40% fruit jellies, 25% chocolates and 35% chews.

Laura's Lollies buys the fruit jellies for \$6 a kilogram, the chocolates for \$10 a kilogram
and the chews for \$5 a kilogram.

a Write a 3 × 3 matrix, M, that represents the percentages of the three types of sweets in
the three mixes.   2 marks

b Write a 3 × 1 matrix, C, that represents the cost per kilogram of the three types of sweets.   1 mark

c If f, c and l are the cost price, in dollars per kilogram, of the Fruit, Chocolate and Lasting
mixes of sweets respectively, write a matrix equation involving M and C that will enable

you to find the matrix, $P = \begin{bmatrix} f \\ c \\ l \end{bmatrix}$.   3 marks

Question 70.2 (6 marks)

At a particular mine, an ore body is being mined and processed for the metals copper, lead, zinc and iron.

The metals in the ore are separated by a process of crushing, grinding and froth flotation into four concentrates, each of which contains a high proportion of one of the metals and traces of the other three metals.

For example, the copper concentrate contains 20% copper, 1% lead, 1% zinc and 5% iron. The following table gives the percentages of each metal in the concentrates.

Metal	Concentrate			
	Copper	Lead	Zinc	Iron
Copper	20	0.25	0.5	0.2
Lead	1	55	1	0.5
Zinc	1	1	45	1
Iron	5	2	2	20

The ore body contains 1% copper, 8% lead, 15% zinc and 10% iron and 2400 tonnes of ore are processed each day. If x, y, z and w are the amounts, in tonnes, of copper, lead, zinc and iron concentrate produced in a day respectively, then the copper balance equation will be $\frac{1}{100}(20x + 0.25y + 0.5z + 0.2w) = 2400 \times 1 \times \frac{1}{100}$, the right-hand side of the equation coming from the fact that the 2400 tonnes of ore contain 1% copper. This equation can be simplified to $20x + 0.25y + 0.5z + 0.2w = 2400 \times 1$.

a Write the other three equations that represent the lead balance, the zinc balance and the iron balance.   3 marks

b Write this set of four equations as a matrix equation.   3 marks

71 Pizza planning (15 marks)

Question 71.1 (5 marks)

Let matrix $A = \begin{bmatrix} 5 & 3 & -2 \\ 4 & 2 & 1 \\ 7 & -1 & 4 \end{bmatrix}$.

a   i Multiply A by the matrix $\begin{bmatrix} 1 \\ 1 \\ 1 \end{bmatrix}$.   2 marks

   ii Explain the effect on the elements of A by this product of matrices.   1 mark

b   i Multiply $\begin{bmatrix} 1 & 1 & 1 \end{bmatrix}$ by A.   1 mark

   ii Explain the effect on the elements of A by this product of matrices.   1 mark

Question 71.2 (5 marks)

A pizza shop makes pizzas in four sizes: small, medium, large and family.

For one particular week, they recorded the sizes of pizzas that they sold each day. The data is recorded in the table below.

Size	Number of pizzas sold						
	Mon	**Tues**	**Wed**	**Thu**	**Fri**	**Sat**	**Sun**
Small	28	35	28	36	37	34	45
Medium	36	37	47	32	36	38	47
Large	41	42	40	35	51	56	61
Family	30	33	39	33	78	76	83

a   Write the information in the table above as a 4×7 matrix, P. 1 mark

b   Write down a suitable matrix, A, such that when P is multiplied by A the resulting matrix will give the number of each size of pizza sold in the week. 1 mark

c   Calculate the matrix PA. 1 mark

d   Write down a suitable matrix, B, such that when P is multiplied by B, the resulting matrix will give the number of pizzas sold on each day of the week. 1 mark

e   Calculate the matrix BP. 1 mark

Question 71.3 (5 marks)

The cost of each of the components in the production of each pizza size, and the selling price, is given in the table below.

Size	Cost of production (cents)				Selling price ($)
	Base	**Topping**	**Cooking**	**Labour**	
Small	32	110	8	200	9.50
Medium	50	175	10	240	11.50
Large	72	250	12	280	13.50
Family	98	345	15	320	17.50

a   Use matrix methods to find the total production cost for each pizza size. Give your answers in cents. 1 mark

b   If *profit = selling price – cost of production*, write a matrix equation that will enable you to find the profit, in dollars correct to the nearest cent, made on each size of pizza. 2 marks

c   Use matrix methods and your answer to Question **71.2** part **c** to find the total profit, in dollars correct to the nearest cent, made in this shop for the week. 2 marks

72 **Playing to win** (15 marks)

Six teams, labelled A, B, C, D, E and F, are planning to play a round-robin competition where each team plays each of the other teams once.

Question 72.1 (1 marks)

State the number of matches that will be played in this competition.   1 mark

Unfortunately, team F had to withdraw after they had only played two matches and the teams that team F did not play were credited with a win for the competition.

The results of the competition are given in the dominance matrix below. A win is represented by a '1' and a loss by a '0'.

$$
M = \begin{array}{c} \\ A \\ B \\ C \\ D \\ E \\ F \end{array}
\begin{array}{c} \text{Defeated} \\ \begin{array}{cccccc} A & B & C & D & E & F \end{array} \\
\left[\begin{array}{cccccc}
0 & 0 & 1 & 0 & 1 & 1 \\
1 & 0 & 1 & 1 & 0 & 0 \\
0 & 0 & 0 & 0 & 1 & 1 \\
1 & 0 & 1 & 0 & 1 & 0 \\
0 & 1 & 0 & 0 & 0 & 1 \\
0 & 1 & 0 & 1 & 0 & 0
\end{array} \right] \end{array}
$$

Question 72.2 (2 marks)

a Which teams did team F play?   1 mark

b Which teams won two matches that they actually played?   1 mark

Question 72.3 (8 marks)

a A dominance vector is a column matrix found by adding all the elements in each of the
    rows of the dominance matrix. Complete the dominance vector, R, for this competition.

$$
R = \begin{bmatrix} 3 \\ \\ \\ \\ \\ \end{bmatrix}
$$
                                                                                                      1 mark

b Use R to state the three teams that are ranked first equally in this round-robin competition. 1 mark

c Use CAS to find the 6×6 matrix M^2.

$$
M^2 = \begin{bmatrix} & & \\ & & \\ & & \\ & & \end{bmatrix}
$$
                                                                                                      2 marks

d What does the matrix M^2 tell us?   1 mark

e Use CAS to find $M + M^2$ and, hence, write the dominance vector associated with $M + M^2$.   2 marks

f Use the dominance vector from part **e** to rank the teams in this competition.   1 mark

Question 72.4 (4 marks)

The teams that played team F thought that they were penalised in the ranking from Question **72.3** part **f**, so it was decided to remove all of the results for team F.

a Write down the revised 5×5 dominance matrix, Q.

$$Q = \begin{bmatrix} 0 & 0 & 1 & 0 & 1 \\ & & & & \\ & & & & \\ & & & & \\ & & & & \end{bmatrix}$$ 1 mark

b Use CAS to find $Q + Q^2$.

$$Q + Q^2 = \begin{bmatrix} & & \\ & & \\ & & \\ & & \end{bmatrix}$$ 1 mark

c Write down the new dominance vector for $Q + Q^2$.   1 mark

d Write down the ranking, first to last, of the five teams.   1 mark

73 Read all about it (12 marks)

A suburban newsagent is asked to collect data about the number of four popular daily newspapers that he sells in his shop on weekdays.

The number of each paper sold and the daily takings are recorded for Monday to Thursday of the first week.

		Newspaper				
		A	**B**	**C**	**D**	**Takings ($)**
Weekday	**Mon**	123	108	48	12	348.00
	Tues	136	b	43	15	377.60
	Wed	115	133	36	11	345.30
	Thurs	145	128	45	18	405.50

Unfortunately, the number of newspaper B sold on Tuesday, b, was lost.

Question 73.1 (5 marks)

a **i** Write the number of each paper sold for the four days as a 4×4 matrix, P.

   ii Write the takings for the four days as a column matrix, Q.   2 marks

b If newspaper A costs \$1.20, newspaper B costs \$1.00, newspaper C costs \$1.30 and newspaper D costs \$2.50, write a suitable matrix X, representing the cost of each of the newspapers, such that $PX = Q$.   1 mark

c Find the missing figure b, the number of newspaper B sold on Tuesday.   2 marks

Question 73.2 (2 marks)

In the second week, the price of two of the newspapers increased.

The newsagent again collected the numbers of papers sold and the takings for Monday to Thursday.

		Newspaper				
		A	**B**	**C**	**D**	**Takings ($)**
Weekday	**Mon**	130	111	45	11	370.85
	Tues	143	126	40	14	403.20
	Wed	121	135	42	12	384.60
	Thurs	136	128	39	14	395.55

The newsagent is a very busy man and he has forgotten which of the newspapers increased in price and by how much.

If newspaper A costs a, newspaper B costs b, newspaper C costs c and newspaper D costs d, write a suitable matrix equation that will enable you to find the values of a, b, c and d.

Question 73.3 (5 marks)

It has been observed that most newspaper buyers do not change the paper that they buy from day-to-day.

The following transition matrix for the newspapers A, B, C and D has been established from records.

$$
\begin{array}{c}
\text{Buy tomorrow}
\end{array}
\quad
\begin{array}{cc}
& \begin{array}{cccc} \text{Buy today} \\ A \quad\ B \quad\ C \quad\ D \end{array} \\
\begin{array}{c} A \\ B \\ C \\ D \end{array} &
\left[\begin{array}{cccc}
0.96 & 0.04 & 0.11 & 0.06 \\
0.01 & 0.95 & 0.03 & 0 \\
0.02 & 0.01 & 0.85 & 0.01 \\
0.01 & 0 & 0.01 & 0.93
\end{array} \right]
\end{array}
$$

a Use the figures for the number of each type of newspaper sold on Thursday of the second week to construct a 4 × 1 initial state matrix.   1 mark

b Use the transition matrix and your initial state matrix to predict the number of each type of newspaper that would be sold on Friday of the second week.   2 marks

c If the type of newspaper bought on a particular day depends entirely on the type bought on the previous day, and the newsagent sells a total of 320 newspapers per day, find the numbers of each type of newspaper that he can expect to sell in the long term.   2 marks

74 Training (12 marks)

Tickets for the train on a small country train line can be bought from a vending machine at the station. There are three types of ticket: Adult, Child and Concession.

Takings from the machine are collected each night and the vending machine records the number of each type of ticket sold. The following table gives the number of tickets sold in a particular week and the takings for each day.

	Adult	Child	Concession	Takings ($)
Mon	68	143	85	921.90
Tues	53	146	72	821.00
Wed	72	139	64	872.30
Thurs	81	155	77	991.90
Fri	78	145	92	994.50

x, y, and z are the cost of an adult, child and concession ticket respectively.

Question 74.1 (3 marks)

a Write an equation involving x, y and z for the takings on Monday.   1 mark

b Write a **matrix** equation involving the ticket cost matrix, $\begin{bmatrix} x \\ y \\ z \end{bmatrix}$, that will enable you to

calculate the values of x, y and z.   2 marks

Question 74.2 (9 marks)

Trains on this line are considered to be fairly reliable; however, records have revealed the following.

If a train is late today, then there is a 95% chance that the train will be on time tomorrow.

However, if a train is on time today, there is a 10% chance that it will be late tomorrow.

a Complete the following table from the information given.   2 marks

		Today	
		On time	Late
Tomorrow	On time		95%
	Late	10%	

b Assuming that the punctuality of a train on a particular day is dependent on its punctuality on the previous day, construct a transition matrix, T, for this situation.   1 mark

c If the train was on time on Monday, write an initial state matrix, S_0.   2 marks

d If the train was on time on Monday, find the chance that the train will be on time on

   i Tuesday   1 mark

   ii Friday.

    Give your answer to one decimal place.   1 mark

e In the long term, what is the chance of a train being on time? Give your answer to one decimal place.   2 marks

75 Island hopping (15 marks)

A cluster of three islands, A, B and C, in an archipelago, is inhabited by people who readily move from one island to another. Some of the people move away from the islands altogether.

The number of people on each of the islands at the beginning of 2016 is given in the matrix

$$S_{2016} = \begin{bmatrix} 232 \\ 375 \\ 483 \\ 0 \end{bmatrix} \begin{matrix} A \\ B \\ C \\ L \end{matrix}$$

The transition diagram below shows the way that people move from one island to another or leave (L) the islands altogether.

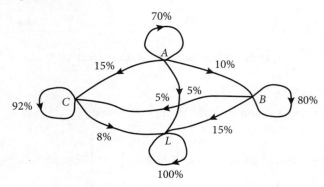

The information in the transition diagram is to be used to write the transition matrix T.

Question 75.1 (9 marks)

a Complete this transition matrix.

3 marks

$$T = \begin{matrix} & \begin{matrix} A & B & C & L \end{matrix} & \\ & \begin{bmatrix} 0.7 & 0 & 0 & 0 \\ \dots & 0.8 & 0 & 0 \\ 0.15 & \dots & 0.92 & \dots \\ 0.05 & 0.15 & 0.08 & 1 \end{bmatrix} & \begin{matrix} A \\ B \\ C \\ L \end{matrix} \end{matrix} \text{ Next year}$$

This year (above the matrix)

b Explain the figure in the third row, third column.

1 mark

If people on the islands continue to move in this way, the matrix S_n will contain the number of inhabitants of island A, island B, island C and the inhabitants who leave the islands (L) at the beginning of the nth year.

c Using the rule $S_{n+1} = TS_n$, find

  i S_{2017}

1 mark

  ii the expected number of people on island A at the beginning of 2018

1 mark

  iii the number of people on island B at the beginning of 2020

1 mark

  iv the total number of people who have left the three islands at the beginning of 2026

1 mark

  v the number of people on island C in the long term.

1 mark

Question 75.2 (6 marks)

The governing body of the three islands does not want them to become uninhabited over the long term, so it has decided to start an incentive scheme for people to move to these islands each year.

Each year, they are offering incentives for 50 people to move to island A, 40 people to island B and 20 people to island C.

a Complete the column matrix Q that represents the number of people moving to the islands under the incentive scheme.

$$Q = \begin{bmatrix} \ldots \\ \ldots \\ \ldots \\ 0 \end{bmatrix} \begin{matrix} A \\ B \\ C \\ L \end{matrix}$$

1 mark

The matrix S_{n+1} now becomes $S_{n+1} = TS_n + Q$.

b Using $S_{2016} = \begin{bmatrix} 232 \\ 375 \\ 483 \\ 0 \end{bmatrix}$ and matrix Q from part **a**, find

 i the expected number of inhabitants of island A in 2017   1 mark

 ii the expected number of inhabitants of island B at the beginning of 2020   1 mark

 iii the total number of inhabitants of the three islands at the beginning of 2020.   1 mark

c Explain what is happening to each of the populations of islands A and C over the first five years of the incentive scheme. Give figures to support your explanation.   2 marks

76 VCAA 2011 Exam 2 (15 marks)

Question 76.1 (3 marks) ©VCAA 2011 2MQ1 ●●

The diagram below shows the feeding paths for insects (I), birds (B) and lizards (L). The matrix E has been constructed to represent the information in this diagram. In matrix E, a '1' is read as 'eat' and a '0' is read as 'do not eat'.

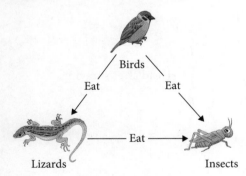

$$E = \begin{bmatrix} 0 & 1 & 1 \\ 0 & 0 & 0 \\ 0 & 1 & 0 \end{bmatrix} \begin{matrix} I \\ B \\ L \end{matrix}$$

with column headers $I \quad B \quad L$

a Referring to insects, birds or lizards

 i what does the '1' in column B, row L, of matrix E indicate?   1 mark

 ii what does the row of zeros in matrix E indicate?   1 mark

The diagram below shows the feeding paths for insects (I), birds (B), lizards (L) and frogs (F).

The matrix Z has been set up to represent the information in this diagram.

Matrix Z has not been completed.

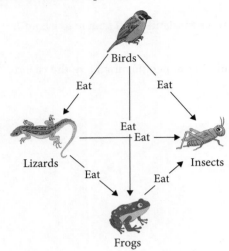

$$Z = \begin{array}{c} \begin{array}{cccc} I & B & L & F \end{array} \\ \left[\begin{array}{cccc} 0 & 1 & 1 & - \\ 0 & 0 & 0 & - \\ 0 & 1 & 0 & - \\ - & - & - & - \end{array} \right] \begin{array}{c} I \\ B \\ L \\ F \end{array} \end{array}$$

b Complete the matrix Z above by writing in the seven missing elements. 1 mark

Question 76.2 (4 marks) ©VCAA 2011 2MQ2 ●●■

To reduce the number of insects in a wetland, the wetland is sprayed with an insecticide.

The numbers of insects (I), birds (B), lizards (L) and frogs (F) in the wetlands that has been sprayed with insecticide are displayed in the matrix N below.

$$N = \begin{array}{c} \begin{array}{cccc} I & B & L & F \end{array} \\ \left[\begin{array}{cccc} 100\,000 & 400 & 1000 & 800 \end{array} \right] \end{array}$$

Unfortunately, the insecticide that is used to kill the insects can also kill birds, lizards and frogs.

The proportions of insects, birds, lizards and frogs that have been killed by the insecticide are displayed in the matrix D below.

Alive before spraying

$$D = \begin{array}{c} \begin{array}{cccc} I & B & L & F \end{array} \\ \left[\begin{array}{cccc} 0.995 & 0 & 0 & 0 \\ 0 & 0.05 & 0 & 0 \\ 0 & 0 & 0.025 & 0 \\ 0 & 0 & 0 & 0.30 \end{array} \right] \begin{array}{c} I \\ B \\ L \\ F \end{array} \end{array} \quad \text{Dead after spraying}$$

a Evaluate the matrix product $K = ND$.

$K =$ 1 mark

b Use the information in matrix K to determine the number of birds that have been killed by the insecticide. 1 mark

c Evaluate the matrix product $M = KF$, where $F = \begin{bmatrix} 0 \\ 1 \\ 1 \\ 1 \end{bmatrix}$.

$M =$ 1 mark

d In the context of the problem, what information does matrix M contain? 1 mark

Question 76.3 (8 marks) ©VCAA 2011 2MQ3 ●●●

A breeding program is started in the wetlands. It is aimed at establishing a colony of native ducks.

The matrix W_0 displays the number of juvenile female ducks (J) and the number of adult female ducks (A) that are introduced to the wetlands at the start of the breeding program.

$$W_0 = \begin{bmatrix} 32 \\ 64 \end{bmatrix} \begin{matrix} J \\ A \end{matrix}$$

a In total, how many female ducks are introduced to the wetlands at the start of the breeding program?

1 mark

The number of juvenile female ducks (J) and the number of adult female ducks (A) in the colony at the end of Year 1 of the breeding program is determined using the matrix equation.

$$W_1 = BW_0$$

In this equation, B is the breeding matrix

$$B = \begin{matrix} J \quad A \end{matrix}$$
$$B = \begin{bmatrix} 0 & 2 \\ 0.25 & 0.5 \end{bmatrix} \begin{matrix} J \\ A \end{matrix}$$

b Determine W_1.

$$W_1 =$$

1 mark

The number of juvenile female ducks (J) and the number of adult female ducks (A) in the colony at the end of Year n of the breeding program is determined using the matrix equation

$$W_n = BW_{n-1}$$

The graph below is incomplete because the points for the end of Year 3 of the breeding program are missing.

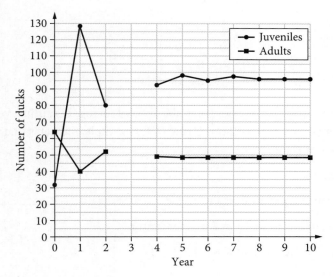

c **i** Use the matrices to calculate the number of juvenile and the number of adult female ducks expected in the colony at the end of Year 3 of the breeding program. Plot the corresponding points on the graph.

2 marks

   ii Use matrices to determine the expected total number of female ducks in the colony in the long term.

1 mark

      Write your answer correct to the nearest whole number.

The breeding matrix B assumes that, on average, each adult female duck lays and hatches two female eggs for each year of the breeding program.

If each adult female duck lays and hatches only one female egg each year, it is expected that the duck colony in the wetland will not be self-sustaining and will, in the long run, die out.

The matrix equation $W_n = PW_{n-1}$ with a different breeding matrix

$$P = \begin{array}{c} \quad J \quad\ A \\ \begin{bmatrix} 0 & 1 \\ 0.25 & 0.5 \end{bmatrix} \begin{array}{c} J \\ A \end{array} \end{array}$$

and the initial state matrix $W_0 = \begin{bmatrix} 32 \\ 64 \end{bmatrix} \begin{array}{c} J \\ A \end{array}$ models this situation.

d During which year of the breeding program will the number of female ducks in the colony halve? 1 mark

Changing the number of juvenile and adult female ducks at the start of the breeding program will also change the expected size of the colony.

e Assuming the same breeding matrix, P, determine the number of juvenile ducks and the number of adult ducks that should be introduced into the program at the beginning so that, at the end of Year 2, there are 100 juvenile female ducks and 50 adult female ducks. 2 marks

77 VCAA 2012 Exam 2 (15 marks)

Question 77.1 (4 marks) ©VCAA 2012 2MQ1 ◐

Matrix F below shows the flight connections for an airline that serves four cities, Anvil (A), Berga (B), Cantor (C), and Dantel (D).

$$F = \begin{array}{c} \qquad\qquad From \\ \quad A \quad B \quad C \quad D \\ \begin{bmatrix} 0 & 1 & 0 & 0 \\ 1 & 0 & 1 & 0 \\ 0 & 0 & 0 & 1 \\ 0 & 1 & 0 & 0 \end{bmatrix} \begin{array}{c} A \\ B \\ C \\ D \end{array} \end{array} \ To$$

In this matrix, the '1' in column C row B, for example, indicates that, using this airline, you can fly directly from Cantor to Berga. The '0' in column C row D, for example, indicates that you cannot fly directly from Cantor to Dantel.

a Complete the following sentence.

On this airline, you can fly directly from Berga to _____ and _____. 1 mark

b List the route that you must follow to fly from Anvil to Cantor. 1 mark

c Evaluate the matrix product $G = KF$, where $K = \begin{bmatrix} 1 & 1 & 1 & 1 \end{bmatrix}$.

$G =$ 1 mark

d In the context of the problem, what information does matrix G contain? 1 mark

Question 77.2 (3 marks) ©VCAA 2012 2MQ2 ●●

Rosa uses the following six-digit pin number for her bank account: 216342

With her knowledge of matrices, she decides to use matrix multiplication to disguise this pin number.

First, she writes the six digits in the 2 × 3 matrix A.

$$A = \begin{bmatrix} 2 & 6 & 4 \\ 1 & 3 & 2 \end{bmatrix}$$

Next, she creates a new matrix by forming the matrix product, $C = BA$,

where $B = \begin{bmatrix} 1 & -1 \\ 2 & -1 \end{bmatrix}$.

a   **i** Determine the matrix $C = BA$.

   $C =$ 1 mark

   ii From the matrix C, Rosa is able to write down a six-digit number that disguises her original pin number. She uses the same pattern that she used to create matrix A from the digits 216342.

   Write down the new six-digit number that Rosa uses to disguise her pin number.   1 mark

b   Show how the original matrix A can be regenerated from matrix C.   1 mark

Question 77.3 (8 marks) ©VCAA 2012 2MQ3 ●●●

When a new industrial site was established at the beginning of 2011, there were 350 staff at the site.

The staff comprised 100 apprentices (A), 200 operators (O) and 50 professionals (P).

At the beginning of each year, staff can choose to stay in the same job, move to a different job at the site or leave the site (L).

The number of staff in each category at the beginning of 2011 is given in the matrix

$$S_{2011} = \begin{bmatrix} 100 \\ 200 \\ 50 \\ 0 \end{bmatrix} \begin{matrix} A \\ O \\ P \\ L \end{matrix}$$

The transition diagram below shows the way in which staff are expected to change their jobs at the site each year.

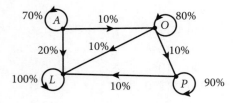

a   How many staff at the site are expected to be working in their same jobs after one year?   1 mark

The information in the transition diagram has been used to write the transition matrix T.

$$\begin{array}{c} \text{This year} \\ \begin{array}{cccc} A & O & P & L \end{array} \\ T = \begin{bmatrix} 0.70 & 0 & 0 & 0 \\ 0.10 & 0.80 & 0 & 0 \\ 0 & 0.10 & 0.90 & 0 \\ 0.20 & 0.10 & 0.10 & 1.00 \end{bmatrix} \begin{array}{c} A \\ O \\ P \\ L \end{array} \text{ Next year} \end{array}$$

b Explain the meaning of the entry in the fourth row and fourth column of transition
matrix T. 1 mark

If staff at the site continue to change their jobs in this way, the matrix S_n will contain the
number of apprentices (A), operators (O), professionals (P) and staff who leave the site (L)
at the beginning of the nth year.

c Using the rule $S_{n+1} = TS_n$, find

   i S_{2012} 1 mark

   ii the expected number of operators at the site at the beginning of 2013 1 mark

   iii the beginning of which year the number of operators at the site first drops below 30 1 mark

   iv the total number of staff at the site in the longer term. 1 mark

Suppose the manager decides to bring 30 new apprentices, 20 new operators and 10 new
professionals to the site at the beginning of each year.

The matrix S_{n+1} will then be given by $S_{n+1} = T S_n + A$ where $S_{2011} = \begin{bmatrix} 100 \\ 200 \\ 50 \\ 0 \end{bmatrix} \begin{array}{c} A \\ O \\ P \\ L \end{array}$ and $A = \begin{bmatrix} 30 \\ 20 \\ 10 \\ 0 \end{bmatrix} \begin{array}{c} A \\ O \\ P \\ L \end{array}$

d Find the expected number of operators at the site at the beginning of 2013. 2 marks

78 VCAA 2014 Exam 2 (15 marks)

Question 78.1 (6 marks) ©VCAA 2014 2MQ1

A small city is divided into four regions: Northern (N), Eastern (E), Southern (S),
and Western (W).

The number of adult males (M) and the number of adult females (F) living in each
of the regions in 2013 is shown in matrix V below.

$$V = \begin{array}{c} \begin{array}{cc} M & F \end{array} \\ \begin{bmatrix} 1360 & 1460 \\ 1680 & 1920 \\ 900 & 1060 \\ 1850 & 1770 \end{bmatrix} \begin{array}{c} N \\ E \\ S \\ W \end{array} \end{array}$$

a Write down the order of matrix V. 1 mark

b How many adult males lived in the Western region in 2013? 1 mark

c In terms of the population of the city, what does the sum of the elements in the second
column of matrix V represent? 1 mark

An election is to be held in the city.

All of the adults in each of the regions of the city will vote in the election.

One of the election candidates, Ms Aboud, estimates that she will receive 45% of the male votes and 55% of the female votes in the election.

This information is shown in matrix P below.

$$P = \begin{bmatrix} 0.45 \\ 0.55 \end{bmatrix} \begin{matrix} M \\ F \end{matrix}$$

d Explain, in terms of rows and columns, why the matrix product $V \times P$ is defined.   1 mark

The product of matrices V and P is shown below.

$$V \times P = \begin{bmatrix} 1360 & 1460 \\ 1680 & 1920 \\ 900 & 1060 \\ 1850 & 1770 \end{bmatrix} \times \begin{bmatrix} 0.45 \\ 0.55 \end{bmatrix} = \begin{bmatrix} w \\ 1812 \\ 988 \\ 1806 \end{bmatrix}$$

e Using appropriate elements from the matrix product $V \times P$, write a calculation to show that the value of w is 1415.   1 mark

f How many votes does Ms Aboud expect to receive in the election?   1 mark

Question 78.2 (6 marks) ©VCAA 2014 2MQ2 ●●

There are three candidates in the election: Ms Aboud (A), Mr Broad (B) and Mr Choi (C).

The election campaign will run for six months, from the start of January until the election at the end of June.

A survey of voters found that voting preference can change from month to month leading up to the election.

The transition diagram below shows the percentages of voters who are expected to change their preferred candidate from month to month.

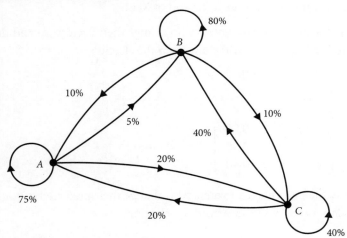

a **i** Of the voters who prefer Mr Choi this month, what percentage are expected to prefer Ms Aboud next month?   1 mark

   ii Of the voters who prefer Ms Aboud this month, what percentage are expected to change their preferred candidate next month?   1 mark

In January, 12 000 voters are expected in the city. The number of voters in the city is expected to remain constant until the election is held in June.

The state matrix that indicates the number of voters who are expected to have a preference to each candidate in January, S_1, is given below.

$$S_1 = \begin{bmatrix} 6000 \\ 3840 \\ 2160 \end{bmatrix} \begin{matrix} A \\ B \\ C \end{matrix}$$

b How many voters are expected to change their preference to Mr Broad in February?   1 mark

The information in the transition diagram has been used to write the transition matrix, T, shown below.

$$\begin{matrix} & \text{This month} \\ & A \quad\;\; B \quad\;\; C \end{matrix}$$

$$T = \begin{bmatrix} 0.75 & 0.10 & 0.20 \\ 0.05 & 0.80 & 0.40 \\ 0.20 & 0.10 & 0.40 \end{bmatrix} \begin{matrix} A \\ B \\ C \end{matrix} \text{ Next month}$$

c   **i** Evaluate the matrix $S_3 = T^2 S_1$ and write it down in the space below.

Write the elements, correct to the nearest whole number.

$$S_3 = \begin{bmatrix} \\ \\ \end{bmatrix}$$ 1 mark

   ii What information does matrix S_3 contain?   1 mark

d Using matrix T, how many votes would the winner of the election in June be expected to receive?

Write your answer, correct to the nearest whole number.   1 mark

Question 78.3 (3 marks) ©VCAA 2014 2MQ3 ●●●

Mr Choi may need to withdraw from the election at the end of May.

Matrix T, shown below, shows the percentage of voters who change their preferred candidate, from month to month, **before** Mr Choi would withdraw from the election.

$$\begin{matrix} & \text{This month} \\ & A \quad\;\; B \quad\;\; C \end{matrix}$$

$$T = \begin{bmatrix} 0.75 & 0.10 & 0.20 \\ 0.05 & 0.80 & 0.40 \\ 0.20 & 0.10 & 0.40 \end{bmatrix} \begin{matrix} A \\ B \\ C \end{matrix} \text{ Next month}$$

Matrix T_1, shown below, shows the percentage of voters who change their preferred candidate, from May to June, **after** Mr Choi would withdraw from the election.

$$\begin{matrix} & \text{May} \\ & A \quad\;\; B \quad\;\; C \end{matrix}$$

$$T_1 = \begin{bmatrix} 0.75 & 0.15 & 0.6 \\ 0.25 & 0.85 & 0.4 \\ 0 & 0 & 0 \end{bmatrix} \begin{matrix} A \\ B \\ C \end{matrix} \text{ June}$$

Consider the voters who preferred Mr Broad in May and who were expected to prefer Mr Choi in June.

a What percentage of these voters are now expected to prefer Mr Broad in June? 1 mark

The state matrix that indicates the number of voters who are expected to have a preference for each candidate in January, S_1, is given below.

$$S_1 = \begin{bmatrix} 6000 \\ 3840 \\ 2160 \end{bmatrix} \begin{matrix} A \\ B \\ C \end{matrix}$$

b If Mr Choi withdraws, how many votes is Mr Broad expected to receive in the election in June?

Write your answer, correct to the nearest vote. 2 marks

79 VCAA 2015 Exam 2 (15 marks)

Question 79.1 (5 marks) ©VCAA 2015 2MQ1

Students in a music school are classified according to three ability levels: beginner (B), intermediate (I) or advanced (A).

Matrix S_0, shown below, lists the number of students at each level in the school for a particular week.

$$S_0 = \begin{bmatrix} 20 \\ 60 \\ 40 \end{bmatrix} \begin{matrix} B \\ I \\ A \end{matrix}$$

a How many students in total are in the music school that week? 1 mark

The music school has four teachers, David (D), Edith (E), Flavio (F) and Geoff (G).

Each teacher will teach a proportion of the students from each level, as shown in matrix P below.

$$\begin{matrix} D & E & F & G \end{matrix}$$
$$P = \begin{bmatrix} 0.25 & 0.5 & 0.15 & 0.1 \end{bmatrix}$$

The matrix product, $Q = S_0 P$, can be used to find the number of students from each level taught by each teacher.

b **i** Complete the matrix Q, shown below, by writing the missing elements in the shaded boxes.

$$Q = \begin{bmatrix} 5 & ■ & 3 & 2 \\ 15 & 30 & ■ & 6 \\ 10 & 20 & 6 & 4 \end{bmatrix}$$ 1 mark

ii How many intermediate students does Edith teach? 1 mark

The music school pays the teachers $15 per week for each beginner student, $25 per week for each intermediate student and $40 per week for each advanced student.

These amounts are shown in matrix C below.

$$C = \begin{bmatrix} \overset{B}{15} & \overset{I}{25} & \overset{A}{40} \end{bmatrix}$$

The amount paid to each teacher each week can be found using a matrix calculation.

c   i Write down a matrix calculation in terms of Q and C that results in a matrix that lists the amount paid to each teacher each week.   1 mark

   ii How much is paid to Geoff each week?   1 mark

Question 79.2 (3 marks) ©VCAA 2015 2MQ2 ●●

The ability level of the students is assessed regularly and classified as beginner (B), intermediate (I) or advanced (A).

After each assessment, students either stay at their current level or progress to a higher level.

Students cannot be assessed at a level that is lower than their current level.

The expected number of students at each level after each assessment can be determined using the transition matrix, T_1, shown below.

$$T_1 = \begin{matrix} & \begin{matrix} \text{Before assessment} \\ B \quad\ I \quad\ A \end{matrix} \\ \begin{bmatrix} 0.50 & 0 & 0 \\ 0.48 & 0.80 & 0 \\ 0.02 & 0.20 & 1 \end{bmatrix} & \begin{matrix} B \\ I \\ A \end{matrix} \text{ After assessment} \end{matrix}$$

a The element in the third row and third column of matrix T_1 is the number 1.

Explain what this tells you about the advanced-level students.   1 mark

Let matrix S_n be a state matrix that lists the number of students at beginner, intermediate and advanced levels after n assessments.

The number of students in the school, immediately before the first assessment of the year, is shown in matrix S_0 below.

$$S_0 = \begin{bmatrix} 20 \\ 60 \\ 40 \end{bmatrix} \begin{matrix} B \\ I \\ A \end{matrix}$$

b   i Write down the matrix S_1 that contains the expected number of students at each level after one assessment.

Write the elements of this matrix correct to the nearest whole number.   1 mark

   ii How many intermediate-level students have become advanced-level students after one assessment?   1 mark

Question 79.3 (7 marks) ©VCAA 2015 2MQ3   ●●

A new model for the number of students in the school after each assessment takes into account the number of students who are expected to leave the school after each assessment.

After each assessment, students are classified as beginner (B), intermediate (I) or advanced (A) or left the school (L).

Let matrix T_2 be the transition matrix for this new model.

Matrix T_2, shown below, contains the percentages of students who are expected to change their ability level or leave the school after each assessment.

$$
\begin{array}{c}
\text{Before assessment}\\
\begin{array}{cccc} B & I & A & L \end{array}\\
T_2 = \left[\begin{array}{cccc}
0.30 & 0 & 0 & 0\\
0.40 & 0.70 & 0 & 0\\
0.05 & 0.20 & 0.75 & 0\\
0.25 & 0.10 & 0.25 & 1
\end{array}\right]\begin{array}{c} B\\ I\\ A\\ L \end{array}\quad \text{After assessment}
\end{array}
$$

a An incomplete transition diagram for matrix T_2 is shown below.

Complete the transition diagram by adding the missing information.   2 marks

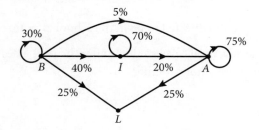

The number of students at each level, immediately before the first assessment of the year, is shown in matrix R_0 below.

$$
R_0 = \left[\begin{array}{c}
20\\
60\\
40\\
0
\end{array}\right]\begin{array}{c} B\\ I\\ A\\ L \end{array}
$$

Matrix T_2, repeated below, contains the percentages of students who are expected to change their ability level or leave the school after each assessment.

$$
\begin{array}{c}
\text{Before assessment}\\
\begin{array}{cccc} B & I & A & L \end{array}\\
T_2 = \left[\begin{array}{cccc}
0.30 & 0 & 0 & 0\\
0.40 & 0.70 & 0 & 0\\
0.05 & 0.20 & 0.75 & 0\\
0.25 & 0.10 & 0.25 & 1
\end{array}\right]\begin{array}{c} B\\ I\\ A\\ L \end{array}\quad \text{After assessment}
\end{array}
$$

b What percentage of students is expected to leave the school after the first assessment?   1 mark

c How many advanced-level students are expected to be in the school after two assessments?

Write your answer correct to the nearest whole number.   1 mark

d After how many assessments is the number of students in the school, correct to the nearest whole number, first expected to drop below 50?   1 mark

Another model for the number of students in the school after each assessment takes into account the number of students who are expected to join the school after each assessment.

Let R_n be the state matrix that contains the number of students in the school immediately after n assessments.

Let V be the matrix that contains the number of students who join the school after each assessment.

Matrix V is shown below.

$$V = \begin{bmatrix} 4 \\ 2 \\ 3 \\ 0 \end{bmatrix} \begin{matrix} B \\ I \\ A \\ L \end{matrix}$$

The expected number of students in the school after n assessments can be determined using the matrix equation $R_{n+1} = T_2 \times R_n + V$ where

$$R_0 = \begin{bmatrix} 20 \\ 60 \\ 40 \\ 0 \end{bmatrix} \begin{matrix} B \\ I \\ A \\ L \end{matrix}$$

e Consider the intermediate-level students expected to be in the school after three assessments.

How many are expected to become advanced-level students after the next assessment?
Write your answer correct to the nearest whole number. 2 marks

Question 80 (8 marks)

The population of females of a small mammal on an island is observed and results are recorded in the table below. These animals rarely live past 4 years so they are grouped in yearly age groups. The birth rate and survival rate for each age group is recorded.

Age group	0–<1	1–<2	2–<3	3–<4
Number of females	32	25	18	14
Birth rate	0	1.1	0.9	0.35
Survival rate	0.48	0.57	0.43	0

The Leslie matrix, L, is given below

$$\begin{bmatrix} 0 & 1.1 & 0.9 & 0.35 \\ 0.48 & 0 & 0 & 0 \\ 0 & 0.57 & 0 & 0 \\ 0 & 0 & 0.43 & 0 \end{bmatrix}$$

a Complete the 4×1 initial state matrix S_0 for this population.   1 mark

b Complete the transition diagram below for this population by filling in the two figures from the table.   2 marks

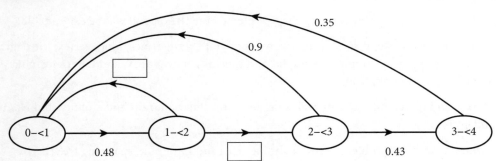

c If $S_{n+1} = LS_n$, find

  i the matrix S_1 1 mark

  ii the number of individuals in the 1–<2 age group after one year   1 mark

  iii the change in the total population after five years.   3 marks

Chapter 4 Networks and decision mathematics
Area of Study 2: Discrete mathematics

Content summary notes

Networks and decision mathematics

Graphs and networks

- Paths, trails and circuits
- Weighted graphs
- Dijkstra's algorithm
- Spanning trees
- Directed graphs and networks
- Reachability
- Network flow
- Critical path analysis
- Assignment problems
- Assignment problems and the Hungarian algorithm

Area of Study 2 Outcome 1

These are the key knowledge points for this chapter, please note that not all will be examinable.

- the conventions, terminology, properties and types of graphs; edge, face, loop, vertex, the degree of a vertex, isomorphic and connected graphs, and the adjacency matrix, and Euler's formula for planar graphs and its application

- the exploring and travelling problem, walks, trails, paths, Eulerian trails and circuits, and Hamiltonian paths and cycles

- the minimum connector problem, trees, spanning trees and minimum spanning trees and Prim's algorithm

- the flow problem, and the minimum cut/maximum flow theorem

- the shortest path problem and Dijkstra's algorithm

- the matching problem and the Hungarian algorithm

- the scheduling problem and critical path analysis

Although you should become familiar with all of the key skills not all the skills are required for the exam. The key skills required for Unit 4 Area of Study 2 are:

- construct graphs, digraphs and networks and their matrix equivalents to model and analyse practical situations

- recognise the exploring and travelling problem and to solve it by utilising the concepts of walks, trails, paths, Eulerian trails and circuits, and Hamiltonian paths and cycles

- recognise the minimum connector problem and solve it by utilising the properties of trees, spanning trees and by determining a minimum spanning tree by inspection or using Prim's algorithm for larger scale problems

- recognise the flow problem, use networks to model flow problems and determine the minimum flow problem by inspection, or by using the minimum cut/maximum flow theorem for larger scale problems

- recognise the shortest path problem and solve it by inspection or using Dijkstra's algorithm for larger scale problems

- recognise the matching problem and solve it by inspection or using the Hungarian algorithm for larger scale problems

- recognise the scheduling problem and solve it by using critical path analysis

VCE Mathematics Study Design 2023–2027 p. 92, © VCAA 2022

4.1   Graphs and networks

- An **undirected graph** has edges without direction arrows.

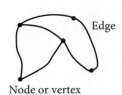

Edge

Node or vertex

- The **degree of a vertex** is the number of edges coming off it. A loop counts as one edge but adds two to the degree of a vertex, so this vertex has degree 3.

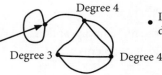

Degree 4

Degree 3   Degree 4

Isolated vertex, degree zero

- A **connected graph** is a graph in which it is possible for one vertex to reach all other vertices by following edges. A minimum of $n - 1$ edges is needed for a graph with n vertices to be connected.

- A **complete graph** is a graph where every vertex is connected to every other vertex. A complete graph with n vertices has $\dfrac{n(n-1)}{2}$ edges.

- A **subgraph** of a graph is part of the graph without the addition of any new edges or vertices. A graph can have many different subgraphs. If an edge is included in the subgraph, then the vertices it is connected to must also be included.

Simple graph   A subgraph

Not a subgraph as there is an edge without its ending

- A **simple graph** has no loops and no multiple edges.

A simple graph

Not simple because there are two edges joining a pair of vertices .

Not simple because there is a loop on one of the vertices.

- A **planar graph** is one that can be drawn so that the edges do not intersect.

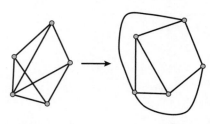

Not planar graph   Planar graph

- **Euler's formula for planar graphs**

  $v + f = e + 2$

  where v is the number of vertices
  $\quad$ e is the number of edges
  $\quad$ f is the number of regions, or faces, that the plane is
  $\quad$ divided into by the graph.

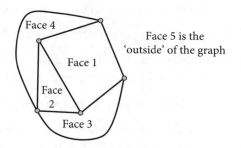

Face 5 is the 'outside' of the graph

The planar graph shown has five vertices ($v = 5$),
eight edges ($e = 8$) and it divides the plane into five
faces ($f = 5$). Confirming Euler's formula for this graph:

$v + f = 5 + 5 = 10$

$e + 2 = 8 + 2 = 10$, so $v + f = e + 2$

A matrix used to show the number of connections on a network graph is called an **adjacency matrix**.

Adjacency matrices can be used to identify networks that have the same structure but look quite different.
Two vertices are adjacent if they are connected by an edge.

Example 1

An adjacency matrix for the network graph is shown below.

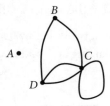

- There are four vertices on the graph, therefore the matrix is a 4 × 4 matrix.

- The graph is **undirected**, so the matrix will be symmetric about the **leading** diagonal.

- There are five edges, so the sum of the elements on one side of the leading diagonal will be 5
  (including the 1 for the loop on the leading diagonal).

The adjacency matrix is:

$$
\begin{array}{c c}
 & \begin{array}{cccc} A & B & C & D \end{array} \\
\begin{array}{c} A \\ B \\ C \\ D \end{array} &
\left[\begin{array}{cccc}
0 & 0 & 0 & 0 \\
0 & 0 & 1 & 1 \\
0 & 1 & 1 & 2 \\
0 & 1 & 2 & 0
\end{array} \right]
\end{array}
$$

The first row and first column
are all zeros. This is because A
is an isolated vertex.

There are two edges
connecting D to C.

There is a '1' in the leading
diagonal. This represents the loop
on vertex C. In this triangle the
numbers add up to 5: the number
of edges.

4.1.1 Paths, trails and circuits

- A **path** is a connected sequence of edges, showing a route that starts at one vertex and ends at another.

- A **circuit** is a path/trail that starts and ends at the same vertex.

- An **Eulerian circuit** is a path/trail that travels once only along every edge of a graph and starts and ends at the same vertex. Vertices can be visited more than once.

  An Eulerian circuit exists on a connected graph if the degree of all the vertices is **even**.

  An Eulerian circuit exists on the network below because each vertex has an even degree. An Eulerian circuit could be *ABCDCADEA*.

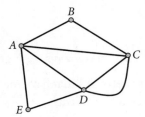

- An **Eulerian trail** travels along every edge of a connected graph but does not start and end at the same vertex.

  An Eulerian trail exists on a connected graph if there are only two vertices that are of odd degree. The Eulerian path will then start and end at the vertices with odd degrees.

  An Eulerian circuit does not exist for the network below because vertices *A* and *D* have degree three (odd); however, an Eulerian trail exists starting at *A* and ending at *D* or vice versa. An Eulerian trail could be *ABCDAED*.

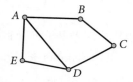

- A **Hamiltonian cycle** is a path that visits each vertex once only and starts and ends on the same vertex. It is not necessary to travel along every edge.

- A **Hamiltonian path** travels through every vertex of a connected graph once but does not start and end at the same vertex.

  A Hamiltonian path for this graph could be *EFABCD*. A Hamiltonian cycle does not exist for this graph.

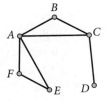

A Hamiltonian cycle for this graph is *ABDCA*. Others are possible.

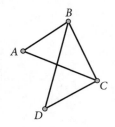

4.1.2 Weighted graphs

To give meaning to many graphs, the edges are given a **weighting**. The weightings can represent distance, time, number of workers, cost, etc.

4.1.3 Dijkstra's algorithm

Dijkstra's algorithm is a systematic method for finding the **shortest path** between two stated vertices on a **weighted graph**.

Using Dijkstra's algorithm, we work from the starting vertex in the direction of the finishing vertex, labelling each vertex as we go with the minimum distance from the start to finish.

Example 2

For the weighted graph shown, find the shortest distance and the path, going from vertex A to vertex G.

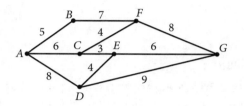

Solution

Starting at vertex A, we find the shortest distance to the connected vertices, i.e., vertices B, C and D.

Vertex B has the shortest distance so we next consider distances from B, i.e., B to F, giving a total of 5 + 7 = 12 for F.

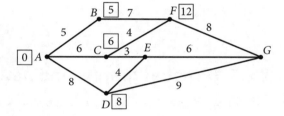

The next shortest distance, not considered, is at vertex C, so we now consider distances from vertex C, i.e., to vertices F and E. This produces two 'shortest distances at F (10 and 12) so the 12 is replaced by 10. Shortest distance to E is 6 + 3 = 9.

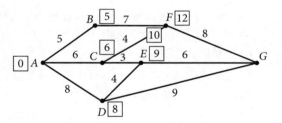

Continue choosing the vertex with shortest distance: Vertex D: D to E gives a distance of 8 + 4 = 12, which is longer than the distance 9 so the 9 remains.

D to G gives the distance 8 + 9 = 17.

E to G gives the distance 15 (less than 17 for D to G).

F to G gives the distance 10 + 8 = 18; more than the distance D to G (15).

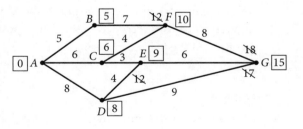

The final shortest distance for each vertex from A is shown on the diagram below.

Hence, the shortest distance from A to G is 15 and this is taking the path ACEG.

4.1.4 Spanning trees

A **spanning tree** is a network in which all vertices are connected and there are no circuits.

A **minimum spanning tree** for a weighted graph is the spanning tree that connects all vertices with the **least possible weight**.

Example 3

Find a minimum spanning tree for the graph.

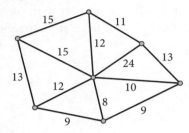

Solution

Using **Prim's algorithm**, starting at the edge with the least weight (8), connect this to another vertex with an edge of least weight. Continue connecting vertices by selecting edges of least weight. The minimum spanning tree is shown in bold.

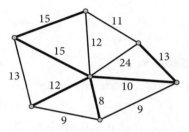

4.1.5 Directed graphs and networks

- A **directed graph**, or **digraph**, is a network graph with arrows to indicate direction on each of the edges.

  The edge joining A to B has an arrow going from A to B, indicating that there is only activity from A to B and not from B to A.

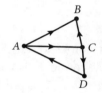

- Weightings, representing quantities such as distance, time, flow, relationships, cost, etc., can also be included on edges.

- An adjacency matrix (sometimes called a connectivity matrix) can be used to store information for a directed network graph. In the graph above, when considering an edge coming from vertex A going to vertex B, a '1' is recorded in the A to B position of the matrix, but a '0' is recorded in the B to A position of the matrix.

Example 4

Find an adjacency or connectivity matrix for the directed graph.

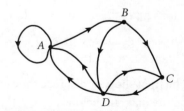

Solution

$$
\begin{array}{c}
 & & \text{To} \\
 & & A \;\; B \;\; C \;\; D \\
\text{From} &
\begin{array}{c}
A \\ B \\ C \\ D
\end{array}
&
\left[
\begin{array}{cccc}
1 & 1 & 0 & 0 \\
0 & 0 & 1 & 1 \\
0 & 0 & 0 & 1 \\
2 & 0 & 1 & 0
\end{array}
\right]
\end{array}
$$

> **Note**
> The matrix is not symmetric about the leading diagonal.
>
> The '1' on the leading diagonal in the A to A position indicates the loop on vertex A.
>
> The sum of the elements in the adjacency matrix is the total number of edges on the graph.

4.1.6 Reachability

In the graph shown, none of the vertices A, C and D is reachable from vertex B as there are no edges leading from B.

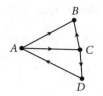

B, however, is reachable from all the other vertices.

Vertex D can reach vertex B via A. This is called a two-stage connection.

If C is the one-stage connectivity matrix, then C^2 will give the two-stage connectivity matrix. For the network graph shown:

There is a two-stage connection from vertex A to vertex D, via C.

$$C = \begin{bmatrix} 0 & 1 & 1 & 0 \\ 0 & 0 & 0 & 0 \\ 0 & 1 & 0 & 1 \\ 1 & 0 & 0 & 0 \end{bmatrix}; \; C^2 = \begin{bmatrix} 0 & 1 & 0 & 1 \\ 0 & 0 & 0 & 0 \\ 1 & 0 & 0 & 0 \\ 0 & 1 & 1 & 0 \end{bmatrix}$$

There is a two-stage connection from vertex D to vertex B, via A.

C^3 will give the three-stage connectivity matrix, C^4 will give the four-stage connectivity matrix, etc.

4.1.7 Network flow

For some weighted digraphs with alternative paths for quantities (fluid, traffic, people, etc.) to flow from a starting vertex to a finishing vertex, it is necessary to establish the maximum flow possible through the network.

- The weighting on an edge is referred to as the **capacity** of the edge.

- The point where the flow starts is often referred to as the **source** and the point where it finishes as the **sink**.

- A **cut** is a line through the graph so that the source is separated from the sink.

- The **capacity of a cut** is the sum of the capacities of the edges that are crossed by the cut; however, if the flow on one of the edges that is cut is **against the flow from source to sink**, then its capacity is not counted in the capacity of the cut.

- A method to find the maximum flow of a system is using the **minimum cut maximum flow** theory. **The cut with the minimum capacity will be the maximum flow for the graph**.

Example 5

Find the maximum flow of the following graph.

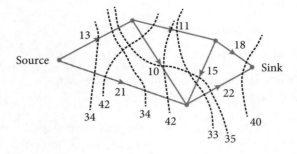

Solution

The maximum flow = the capacity of the minimum cut = 33; the edge with capacity 15 is not counted as it is against the flow.

4.1.8 Critical path analysis

- **Activity tables** and network diagrams for projects.

Example 6

The activities involved in a project are given in the activity table below.

Activity	Duration (days)	Immediate predecessors
A	2	–
B	4	A
C	5	A
D	8	B, C
E	6	B, C
F	3	D, E

The **network diagram for the project** is given below.

The activity, and the duration of the activity, are written on the edges of the graph.

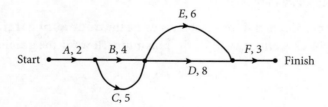

- The **critical path**, from the start to the finish of a network, is the sequence of activities that is the **longest path through the network**.

- A delay in any of the activities on the critical path would cause a delay in the whole project.

- For projects with complex networks, there is a method for identifying the critical path(s) and the **earliest start time (EST)** and **latest start time (LST)** for each of the activities. This process involves what is called **forward and backward scanning**.

EST | LST on the diagram indicate the earliest start time (EST) and latest start time (LST) for each of the activities.

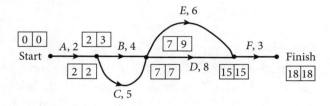

- The critical path is the path through the network along the activities that have the same values for their EST and LST.

For the example, the critical path is *ACDF*. The earliest finishing time for the project is 18 days so this is the longest path through the network.

- **Float time = latest start time − earliest start time**. For the example, the float time for activity *E* is $9 − 7 = 2$ days. Activities on the critical path have a float time of zero.

Crashing a project

It is often possible to decrease the overall time to complete a project by decreasing the duration of some of the activities. This usually happens at a cost and the process is called **crashing the project**. Crashing a project is often a trial-and-error process and the following should be noted.

i If it is possible to reduce the duration of several activities, then it is important to first consider **reducing the duration of activities on the critical path** as these activities determine the time of completion of the project.

ii Reducing the duration of an activity that is not on the critical path may not reduce the project completion time.

iii The process of crashing a project can change the critical path of a project.

4.1.9 Assignment problems

Bipartite graphs

Bipartite graphs are directed or undirected graphs that are used to represent a relationship between two distinct sets.

Example 7

Four people can do a variety of tasks:

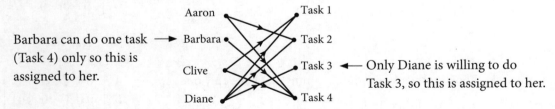

Barbara can do one task → (Task 4) only so this is assigned to her.

Only Diane is willing to do ← Task 3, so this is assigned to her.

The only possible assignment is given below.

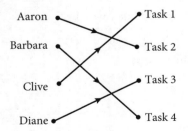

4.1.10 Assignment problems and the Hungarian algorithm

When assignment problems involve weightings, we can use the **Hungarian algorithm** to solve the problem.

Example 8

Operators A, B, C and D are each to be allocated one of four tasks, 1, 2, 3 and 4. Their estimates of the time, in hours, that each of these tasks will take them are summarised in the matrix shown. How should the tasks be assigned so that the total time to complete all four tasks is a minimum?

$$\text{Operator} \quad \begin{array}{c} A \\ B \\ C \\ D \end{array} \begin{bmatrix} 3 & 6 & 5 & 7 \\ 4 & 6 & 6 & 5 \\ 3 & 6 & 7 & 6 \\ 5 & 5 & 6 & 6 \end{bmatrix}$$

with Task columns labelled 1 2 3 4.

Solution

Use the Hungarian algorithm to assign the tasks.

A **minimum of four lines** is needed before we can allocate the task because this is a 4 × 4 matrix.

Step 1

Subtract the minimum value for a row from all elements in the row, then cover the zeros with a minimum number of lines. Only two are needed, as shown.

$$\begin{bmatrix} 0 & 3 & 2 & 4 \\ 0 & 2 & 2 & 1 \\ 0 & 3 & 4 & 3 \\ 0 & 0 & 1 & 1 \end{bmatrix}$$

Step 2

Subtract the minimum value from each of the columns.

Cover the zeros with lines: only three lines are used, as shown.

However, four lines are required, so proceed to step three.

$$\begin{bmatrix} 0 & 3 & 1 & 3 \\ 0 & 2 & 1 & 0 \\ 0 & 3 & 3 & 2 \\ 0 & 0 & 0 & 0 \end{bmatrix}$$

Step 3

Add the minimum uncovered value (1) to each element covered by two lines and subtract it from each uncovered element.

$$\begin{bmatrix} 0 & 2 & 0 & 3 \\ 0 & 1 & 0 & 0 \\ 0 & 2 & 2 & 2 \\ 1 & 0 & 0 & 1 \end{bmatrix}$$

Cover the zeros with the minimum number of lines. Four lines are now needed so the tasks can be allocated.

$$\begin{bmatrix} 0 & 2 & 0 & 3 \\ 0 & 1 & 0 & 0 \\ 0 & 2 & 2 & 2 \\ 1 & 0 & 0 & 1 \end{bmatrix}$$

Step 4

Allocate the tasks. Look for the rows and/or columns that have only one zero.

Column 2 has one zero, so allocate Task 2 to D. Column 4 has one zero, so allocate Task 4 to B.

Row 3 has one zero, so allocate Task 1 to C. This leaves Task 3 allocated to A.

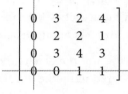

$$\begin{array}{c} A \\ B \\ C \\ D \end{array} \begin{bmatrix} 0 & 2 & \boxed{0} & 3 \\ 0 & 1 & 0 & \boxed{0} \\ \boxed{0} & 2 & 2 & 2 \\ 1 & \boxed{0} & 0 & 1 \end{bmatrix}$$

with Task columns 1 2 3 4.

The minimum time to complete all the tasks is:
$5 + 5 + 3 + 5 = 18$ hours

$$\text{Operator} \quad \begin{array}{c} A \\ B \\ C \\ D \end{array} \begin{bmatrix} 3 & 6 & \boxed{5} & 7 \\ 4 & 6 & 6 & \boxed{5} \\ \boxed{3} & 6 & 7 & 6 \\ 5 & \boxed{5} & 6 & 6 \end{bmatrix}$$

Glossary

adjacency matrices A matrix that shows the number of edges between vertices on a graph. Can be used to identify networks that have the same structure but look quite different.

bipartite graphs Directed or undirected graphs that are used to represent a relationship between two distinct sets.

capacity (of the edge) Referred to as the **weighting** on the edge.
See also **weighting**.

circuit A path/trail that starts and ends at the same vertex.

complete graph A graph where every vertex is connected to every other vertex. A complete graph with n vertices has $\frac{n(n-1)}{2}$ edges.

connected graph A graph in which it is possible for one vertex to reach all other vertices by following edges. A minimum of $n-1$ edges is needed for a graph with n vertices to be connected.

crashing a project It is often possible to decrease the overall time to complete a project by decreasing the duration of some of the activities. This usually happens at a cost and the process is called *crashing the project*.

critical path analysis A step-by-step project management technique that is used to examine every activity in a project and how each affects the project completion time.

degree of a vertex The number of edges, coming off it.

Dijkstra's algorithm A systematic method for finding the shortest path between two stated vertices on a weighted graph.

digraph A network graph with arrows to indicate direction on each of the edges.
See also **directed graph**.

directed graph A network graph with arrows to indicate direction on each of the edges.
See also **digraph**.

Euler circuit A path/trail that travels once only along every edge of a graph and which starts and ends at the same vertex. Vertices can be visited more than once. Exists on a connected graph if the degree of all the vertices is even.

A+ DIGITAL
Revise this topic's key terms and concepts by scanning the QR code or typing the URL into your browser.

https://get.ga/aplus-vcegeneral-maths

Euler's formula for planar graphs $v + f = e + 2$ where v is the number of vertices, e is the number of edges and f is the number of **faces** that the plane is divided into by the graph.

Euler trail A path that travels along every edge of a connected graph but does not start and end at the same vertex. Exists on a connected graph if there are **only two vertices that are of odd degree**. The Euler trail will then start and end at the vertices with odd degrees.

Hamiltonian cycle A path that visits **each vertex once only** and starts and ends on the same vertex. It is not necessary to travel along every edge.

Hamiltonian path A path that travels through every vertex of a connected graph once but does not start and end at the same vertex.

Hungarian algorithm A method used to find the optimum allocation for the assignment problem. When assignment problems also involve weightings, we can use the Hungarian algorithm to solve the problem.

minimum spanning tree (for a weighted graph) The spanning tree that connects all vertices with the **least possible weight**.

network flow For some weighted digraphs with alternative paths for quantities (fluid, traffic, people, etc.) to flow from a starting vertex to a finishing vertex, it is necessary to establish the flow possible through the network.

path A connected sequence of edges, showing a route that starts at one vertex and ends at another.

planar graph A graph that can be drawn so that the edges do not intersect.

Prim's algorithm A method used to determine the minimum spanning tree.

simple graph A graph with no loops and no multiple edges.

Not simple because there are two edges joining a pair of vertices

Not simple because there is a loop on one of the vertices.

spanning tree A network in which all vertices are connected and there are no circuits.

subgraph Part of the graph without the addition of any new edges or vertices.

undirected graph A graph that has edges without direction arrows.

weighted graphs A graph whose edges are labelled with numbers representing physical quantities such as distance, time or cost.

weighting A number attached to an edge on a graph that represents a physical quantity such as distance, time or cost. Also referred to as **capacity** of the edge.

Exam practice

Multiple-choice questions

Undirected graphs and applications: 36 questions

Solutions to this section start on page 280.

Question 1

The most common degree of the vertices in the following network is

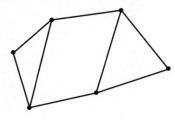

A 0   **B** 1   **C** 2   **D** 3   **E** 4

Question 2

Consider the following graph.

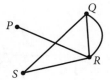

This graph above could be described as

A directed   **B** planar   **C** weighted   **D** complete   **E** simple

Question 3

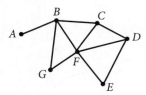

Which one of the following is a spanning tree for the graph above?

A 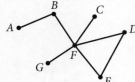   **B** 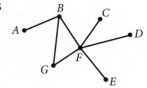   **C**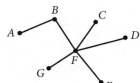

D 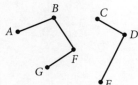   **E**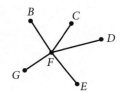

Question 4

For which one of the following graphs would it **not** be possible to apply Euler's formula $v + f = e + 2$?

A   B   C

D   E

Question 5

The planar graph is divided into a number of faces.

The number of faces is

A 2   B 3   C 4   D 5   E 6

Question 6

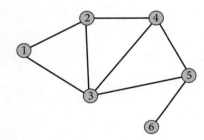

Which one of the following is a subgraph of the graph above?

A 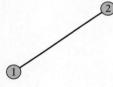   B 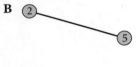   C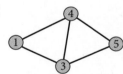

D 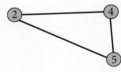   E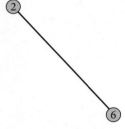

Question 7 ⬤⚫⚫

A city block is occupied by five businesses, *P*, *Q*, *R*, *S* and *T* and the land is divided as shown in the diagram below.

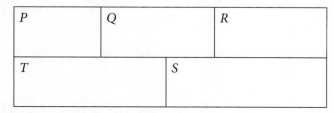

A network diagram where the vertices are the businesses and the edges represent 'shares a boundary with' is given by

A
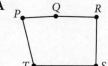

B

C
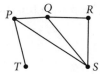

D
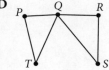

E
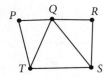

Question 8 ⬤⚫⚫

The number of vertices for a graph with 10 edges is

A 4   **B** 5   **C** 9   **D** 10   **E** 45

Question 9 ⬤⬤⚫

The minimum number of edges required for a graph with *n* vertices to be connected is

A n **B** $n-1$ **C** $n+1$ **D** $n(n-1)$ **E** $\dfrac{n(n-1)}{2}$

Question 10 ⬤⬤⚫

A connected planar graph has twice as many vertices as faces. Given that there are 16 edges in this graph, the number of faces must be

A 6   **B** 8   **C** 12   **D** 16   **E** 32

Question 11 ⬤⚫⚫

Graphs *A*, *B*, *C*, *D* and *E* are planar graphs that each contain four vertices. The degrees of the vertices for each graph are as follows:

Graph *A*:   1   2   3   4
Graph *B*:   2   3   4   5
Graph *C*:   2   3   3   4
Graph *D*:   2   2   4   2
Graph *E*:   2   3   4   4

Which one of the five graphs would contain an Eulerian circuit?

A Graph *A*   **B** Graph *B*   **C** Graph *C*   **D** Graph *D*   **E** Graph *E*

Question 12 🔘🔘⚫

A simple graph has 7 edges. The sum of the degrees of the vertices for this graph is

A 6   **B** 7   **C** 12   **D** 14   **E** 21

Question 13 🔘🔘⚫

The following adjacency matrix is a representation of a graph.

$$\begin{bmatrix} 0 & 2 & 1 & 1 \\ 2 & 0 & 1 & 2 \\ 1 & 1 & 0 & 1 \\ 1 & 2 & 1 & 2 \end{bmatrix}$$

The number of loops that this graph contains is

A 0   **B** 1   **C** 2   **D** 3   **E** 4

Question 14 🔘🔘⚫

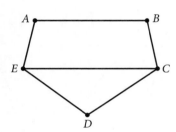

The graph above can be described as

A complete and planar   **B** connected and planar   **C** simple and complete

D planar with no circuits   **E** a tree with one circuit

Question 15 🔘🔘⚫

The minimum number of edges for a graph with ten vertices to be connected is

A 9   **B** 10   **C** 11   **D** 45   **E** 90

Question 16 🔘🔘⚫

A connected planar graph with 10 vertices has 10 faces. The number of edges connecting the vertices in this graph is

A 4   **B** 6   **C** 14   **D** 16   **E** 18

Question 17 🔘🔘⚫

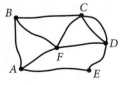

For the graph given above, it is possible to form an Eulerian trail that

A starts at *A* and finishes at *B*.   **B** starts at *A* and finishes at *E*.   **C** starts at *B* and finishes at *C*.

D starts at *E* and finishes at *F*.   **E** starts at *E* and finishes at *B*.

Question 18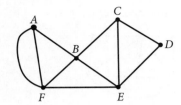

For the graph above, which one of the following is a Hamiltonian cycle?

A *ABCDEF*

B *ABCDEFBFAFEC*

C *FBEDCBAF*

D *EDCBAFE*

E *AFBEDCBA*

Question 19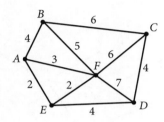

Which of the following is the minimum spanning tree for the graph above?

A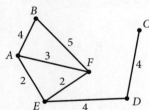

B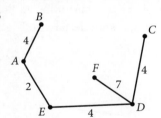

C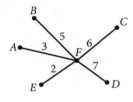

D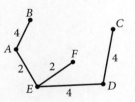

E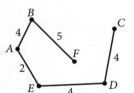

Question 20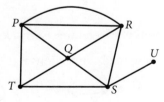

To convert the network above to one that has an Eulerian circuit, we could add edge

A *PT*

B *RU*

C *TU*

D *TS*

E *QU*

Question 21

Which one of the following graphs contains an Euler circuit?

A   B 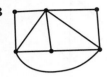   C

D   E

Question 22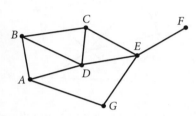

The addition of another edge to the graph above will mean that an Eulerian trail is available. Adding which one of the following edges creates an Eulerian trail?

A *FG*   **B** *DG*   **C** *BE*   **D** *AE*   **E** *AC*

Question 23

The sum of the degrees of all the vertices in the graph above is

A 6   **B** 9   **C** 15   **D** 16   **E** 17

Question 24 ©VCAA 2013 1NQ6

The map above shows the road connections between three towns, *P*, *Q* and *R*.

The graph that could be used to model these road connections is

A   B   C   D   E

Question 25

A connected planar graph has 23 vertices and 36 edges. The number of faces that the plane is divided into by this graph is

A 13   **B** 15   **C** 29   **D** 57   **E** 61

Question 26 ©VCAA 2010 1NQ5 ●●

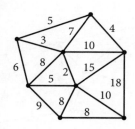

For the network above, the length of the minimal spanning tree is

A 30   **B** 31   **C** 35   **D** 39   **E** 45

Question 27 ●● ■

Which one of the following is **not** a planar graph?

A   **B**   **C**

D   **E**

Question 28 ●● ■

The adjacency matrix for a network graph is given.

$$\begin{bmatrix} 0 & 0 & 2 & 0 \\ 0 & 1 & 1 & 0 \\ 2 & 1 & 0 & 0 \\ 0 & 0 & 0 & 0 \end{bmatrix}$$

Which one of the following is **not** true for the network diagram?

A There are 4 vertices on the diagram.   **B** There are 4 edges on the diagram.

C There is one loop on the diagram.   **D** One vertex is isolated.

E The sum of the degrees of the vertices is 4.

Question 29 ●●●

The length of the minimal spanning tree for the following network is 24.

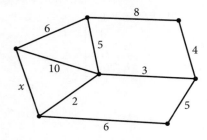

If one of the edges in the minimum spanning tree is labelled as x, then the weight of the edge labelled x must be

A 3   **B** 4   **C** 5   **D** 6   **E** 7

Question 30 ●●●

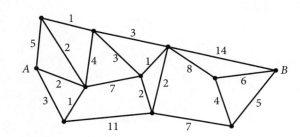

Consider the above graph.

Using Dijkstra's algorithm, or otherwise, the shortest path from A to B has a length of

A 21   **B** 22   **C** 23   **D** 24   **E** 25

Question 31 ●●●

Which one of the following is an adjacency matrix for the graph above?

A
$$\begin{bmatrix} 1 & 2 & 2 & 0 \\ 2 & 0 & 2 & 0 \\ 2 & 2 & 0 & 1 \\ 0 & 0 & 1 & 0 \end{bmatrix}$$

B
$$\begin{bmatrix} 5 & 2 & 1 & 0 \\ 2 & 4 & 2 & 1 \\ 1 & 2 & 4 & 1 \\ 0 & 1 & 1 & 1 \end{bmatrix}$$

C
$$\begin{bmatrix} 1 & 2 & 1 & 0 \\ 2 & 0 & 2 & 0 \\ 1 & 2 & 0 & 1 \\ 0 & 0 & 1 & 0 \end{bmatrix}$$

D
$$\begin{bmatrix} 2 & 3 & 2 & 0 \\ 3 & 0 & 2 & 1 \\ 2 & 2 & 0 & 1 \\ 0 & 1 & 1 & 0 \end{bmatrix}$$

E
$$\begin{bmatrix} 2 & 2 & 1 & 0 \\ 0 & 0 & 2 & 0 \\ 0 & 0 & 0 & 1 \\ 0 & 0 & 0 & 0 \end{bmatrix}$$

Question 32 ●●●

Which one of the following adjacency matrices represents a connected graph?

A
$$\begin{bmatrix} 0 & 2 & 1 & 0 \\ 2 & 0 & 1 & 0 \\ 1 & 1 & 0 & 0 \\ 0 & 0 & 0 & 0 \end{bmatrix}$$

B
$$\begin{bmatrix} 0 & 2 & 1 & 0 \\ 2 & 0 & 1 & 2 \\ 1 & 1 & 0 & 0 \\ 0 & 2 & 0 & 0 \end{bmatrix}$$

C
$$\begin{bmatrix} 0 & 0 & 1 & 0 \\ 0 & 0 & 1 & 0 \\ 1 & 1 & 0 & 0 \\ 0 & 0 & 0 & 1 \end{bmatrix}$$

D
$$\begin{bmatrix} 0 & 0 & 2 & 0 \\ 0 & 0 & 0 & 1 \\ 2 & 0 & 0 & 0 \\ 0 & 1 & 0 & 1 \end{bmatrix}$$

E
$$\begin{bmatrix} 0 & 2 & 0 & 0 \\ 2 & 1 & 0 & 0 \\ 0 & 0 & 1 & 1 \\ 0 & 0 & 1 & 0 \end{bmatrix}$$

Question 33 ⬤⬤⬤

The length of the minimum spanning tree for the network is

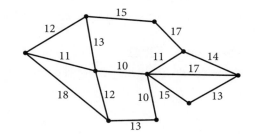

A 93

B 95

C 106

D 108

E 110

Question 34 ©VCAA 2011 1NQ9 ⬤⬤⬤

An Eulerian trail through a network commences at vertex P and ends at vertex Q.

Consider the following five statements about this Eulerian trail and network.

- In the network, there could be three vertices with degree equal to one.

- The trail could have passed through an isolated vertex.

- The trail could have included vertex Q more than once.

- The sum of the degrees of vertices P and Q could equal seven.

- The sum of the degrees of all vertices in the network could equal seven.

How many of these statements are true?

A 0   **B** 1   **C** 2   **D** 3   **E** 4

Question 35 ©VCAA 2009 1NQ8 ⬤⬤⬤

An undirected connected graph has five vertices.

Three of these vertices are of even degree and two of these vertices are of odd degree.

One extra edge is added. It joins two of the existing vertices.

In the resulting graph, it is **not** possible to have five vertices that are

A all of even degree.

B all of equal degree.

C one of even degree and four of odd degree.

D three of even degree and two of odd degree.

E four of even degree and one of odd degree.

Question 36 ⬤⬤⬤

The network diagram below shows a group of buildings and pathways between these buildings.
The weightings on the edges represent the distance, in metres, along the pathways. A security guard
has to walk along all the pathways between the buildings P, Q, R, S and T.

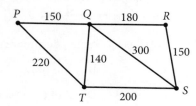

If the security guard must start and finish at building T, then the shortest distance he can travel will be

A 1340 m   **B** 1480 m   **C** 1540 m   **D** 1560 m   **E** 1640 m

Directed graphs and networks: 27 questions

Solutions to this section start on page 284.

Question 37 ◐

Consider the following directed graph.

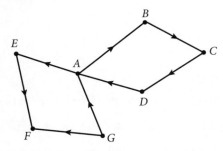

Which vertex cannot be reached from *B*?

A *A*   **B** *D*   **C** *E*   **D** *F*   **E** *G*

Question 38 ◑

The bipartite graph below shows the sports played by four people.

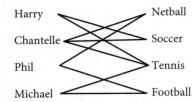

Which one of the following statements follows from the graph?

A All people participate in the same number of sports.

B Phil only plays football.

C Netball is the least chosen sport of these people.

D Chantelle plays a greater variety of sports than the other people.

E Two of these people play both soccer and football.

Question 39 ©VCAA 2011 1NQ2 ◐

The graph below shows the one-step dominances between four farm dogs, Kip, Lab, Max and Nim. In this graph, an arrow from Lab to Kip indicates that Lab has a one-step dominance over Kip.

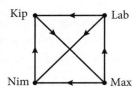

From this graph, it can be concluded that Kip has a two-step dominance over

A Max only.   **B** Nim only.

C Lab and Nim only.   **D** all of the other three dogs.

E none of the other three dogs.

The following information relates to Questions 40 and 41.

The earliest and latest start times (in hours) for eight activities in a project are given in the table below.

Activity	EST	LST
A	0	3
B	0	0
C	5	5
D	8	8
E	6	8
F	7	12
G	16	16
H	20	20

Question 40

The activities that lie on the critical path are

A *ABCGH*   **B** *AEFGH*   **C** *BCDGH*   **D** *ABEFH*   **E** *BCDEFH*

Question 41

The float time, in hours, for activity *F* is

A 2   **B** 3   **C** 4   **D** 5   **E** 6

Question 42

Which one of the following statements is **true** regarding the critical path of a network project?

A It is the shortest path from the beginning to the end of the project.

B It is the longest path from the beginning to the end of the project.

C The activities along the critical path can be delayed without delaying the project.

D The activities along the critical path must all start at the same time.

E The length of the critical path gives the maximum time for project completion.

Question 43

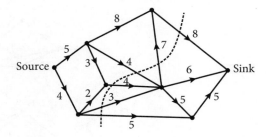

For the graph above, the capacity of the cut shown is

A 16   **B** 17   **C** 20   **D** 24   **E** 31

Question 44 🔘🔘

The following graph indicates a friendship group of six people who occasionally drive each other to work. On this graph, an arrow going from Joe to Ben indicates that Joe has driven Ben to work.

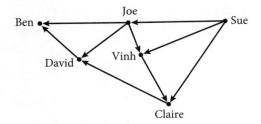

Based on the information contained in this graph, which of the following statements is **not** true?

A Ben has not driven any of these friends to work.

B Sue has not been driven to work by any of these friends.

C David has been driven to work by Claire and Joe.

D Joe has given everyone but Claire a lift to work.

E Vinh has only driven Claire to work.

Question 45 🔘🔘

A precedence table for a project of seven activities is given below.

Activity	Immediate predecessors
A	–
B	A
C	A
D	B
E	C, D
F	B
G	E, F

Which one of the following networks represents this project?

A

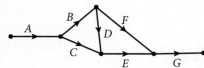

B

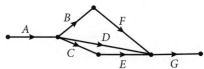

C

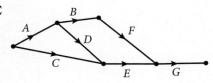

D

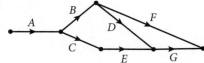

E

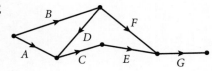

The following information relates to Questions 46, 47 and 48.

The network diagram for a project is given below. The activities and duration (days) are marked on the edges.

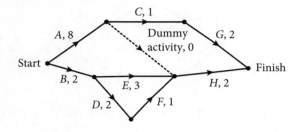

Question 46 ●●●

All the activities that are to be completed before activity *H* can start are

A *A, B, C, D, E, F* and *G*.   **B** *E* and *F*.   **C** *A, B, D, E* and *F*.

D *B, D, E* and *F*.   **E** *A, B* and *E*.

Question 47 ●●●

The latest start time for activity *E* will be

A 4   **B** 5   **C** 6   **D** 7   **E** 8

Question 48 ●●●

The critical path for the project is

A *ACG*   **B** *BEH*   **C** *AEH*   **D** *BDFH*   **E** *AH*

Question 49 ©VCAA 2013 1NQ4 ●●●

Kate, Lexie, Mei and Nasim enter a competition as a team. In this competition, the team must complete four tasks, *W, X, Y* and *Z*, as quickly as possible.

The table shows the time, in minutes, that each person would take to complete each of the four tasks.

	Kate	Lexie	Mei	Nasim
W	6	3	4	6
X	4	3	5	5
Y	5	7	9	6
Z	3	2	5	2

If each team member is allocated one task only, the minimum time in which this team would complete the four tasks is

A 10 minutes   **B** 12 minutes   **C** 13 minutes

D 14 minutes   **E** 15 minutes

Question 50 ©VCAA 2012 1NQ6 ●●

In the digraph, all vertices are reachable from every other vertex.

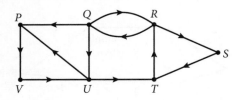

All vertices would still be reachable from every other vertex if we remove the edge in the direction from

A Q to U **B** R to S **C** S to T **D** T to R **E** V to U

The following information relates to Questions 53 and 54.

A connectivity matrix for a network with four nodes is given below.

$$
\begin{array}{c}
\text{To} \\
\begin{array}{cccc}
A & B & C & D
\end{array} \\
\text{From} \quad
\begin{array}{c}
A \\ B \\ C \\ D
\end{array}
\begin{bmatrix}
0 & 1 & 1 & 0 \\
0 & 0 & 1 & 0 \\
1 & 0 & 0 & 1 \\
0 & 1 & 1 & 0
\end{bmatrix}
\end{array}
$$

Question 51 ●●

Which one of the following would be the network represented by this matrix?

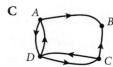

Question 52 ●●

The number of two-step connections from vertex C to vertex B is

A 0 **B** 1 **C** 2 **D** 3 **E** 4

Question 53 ●●

Four tasks, 1, 2, 3 and 4 are to be allocated to four people, A, B, C and D. The capability of these people to do the tasks is given in the graph.

The only possible allocation of tasks to people is

A 1 to A; 2 to C; 3 to B; 4 to D **B** 1 to B; 2 to C; 3 to A; 4 to D

C 1 to A; 2 to C; 3 to D; 4 to B **D** 1 to B; 2 to D; 3 to A; 4 to C

E 1 to A; 2 to D; 3 to C; 4 to B

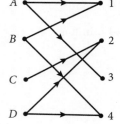

Question 54 ●●●

Four co-workers, Amy, Bill, Claire and Darren are each to be assigned one of four tasks. The time that each take to perform the task (in minutes) is recorded in the following table.

	Task 1	Task 2	Task 3	Task 4
Amy	6	4	9	3
Bill	7	3	10	2
Claire	8	6	11	4
Darren	6	7	14	3

If each person is assigned a different task, then the overall time for all tasks to be completed can be minimised by assigning Claire to

A Task 1 　　　　**B** Task 2 　　　　**C** Task 3 　　　　**D** Task 4 　　　　**E** either Task 1 or Task 3

Question 55 ●●●

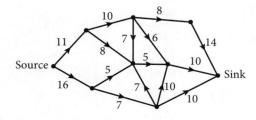

The maximum flow through the network above is

A 21 　　　　**B** 22 　　　　**C** 23 　　　　**D** 26 　　　　**E** 27

Question 56 ●●●

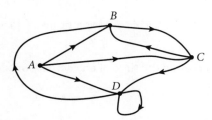

Which one of the following adjacency matrices would represent the graph above?

A $\begin{bmatrix} 0 & 1 & 1 & 1 \\ 1 & 0 & 1 & 1 \\ 1 & 1 & 0 & 1 \\ 1 & 1 & 0 & 1 \end{bmatrix}$ 　　**B** $\begin{bmatrix} 0 & 1 & 1 & 1 \\ 0 & 0 & 1 & 0 \\ 0 & 1 & 0 & 1 \\ 0 & 1 & 0 & 1 \end{bmatrix}$ 　　**C** $\begin{bmatrix} 1 & 1 & 1 & 1 \\ 0 & 1 & 1 & 0 \\ 1 & 1 & 1 & 1 \\ 0 & 1 & 0 & 1 \end{bmatrix}$

D $\begin{bmatrix} 1 & 0 & 1 & 1 \\ 0 & 1 & 1 & 0 \\ 1 & 1 & 0 & 1 \\ 0 & 1 & 1 & 1 \end{bmatrix}$ 　　**E** $\begin{bmatrix} 0 & 1 & 1 & 1 \\ 0 & 0 & 1 & 0 \\ 1 & 1 & 0 & 1 \\ 1 & 1 & 1 & 1 \end{bmatrix}$

Question 57 ●●●

Four people are to be assigned one of four tasks and they are asked to give the time taken to complete each task. These times, in hours, are given in the table.

	Task			
	P	**Q**	**R**	**S**
Peter	3	6	8	5
Jane	3	4	9	5
Rebecca	7	4	10	4
Vince	6	6	6	4

What is the optimal assignment that will minimise the total time for the completion of the four tasks?

A P to Peter, Q to Jane, R to Rebecca, S to Vince

B P to Peter, Q to Jane, S to Rebecca, R to Vince

C S to Peter, Q to Jane, P to Rebecca, Q to Vince

D Q to Peter, P to Jane, R to Rebecca, S to Vince

E Q to Peter, S to Jane, R to Rebecca, S to Vince

Question 58 ©VCAA 2014 1NQ8 ●●●

Which one of the following statements about critical paths is **true**?

A There can only be one critical path in a project.

B A critical path always includes at least two activities.

C A critical path will always include the activity that takes the longest time to complete.

D Reducing the time of any activity on a critical path for a project will always reduce the minimum completion time for the project.

E If there are no other changes, increasing the time of any activity on a critical path will always increase the completion time of a project.

Question 59 ●●●

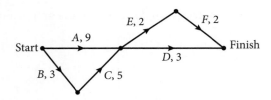

The directed graph above shows the activities and their respective duration (in days) for completing a project. What is the latest time that task D can start in order not to delay the project?

A 1 day after the start   **B** 8 days after the start   **C** 9 days after the start

D 10 days after the start   **E** 13 days after the start

Question 60 ©VCAA 2012 1NQ8 ●●●

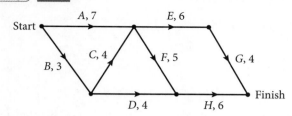

Eight activities, *A*, *B*, *C*, *D*, *E*, *F*, *G* and *H*, must be completed for a project.

The graph above shows these activities and their usual duration in hours.

The duration of each activity can be reduced by one hour.

To complete this project in 16 hours, the minimum number of activities that must be reduced by one hour each is

A 1   **B** 2   **C** 3   **D** 4   **E** 5

Question 61 ●●●

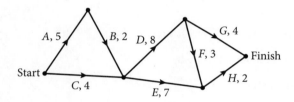

The directed graph above shows the activities and their respective duration (in days) for completing a project. The float time, in days, for activity *E* will be

A 1   **B** 2   **C** 3   **D** 4   **E** 5

Question 62 ●●●

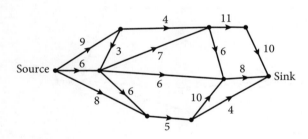

The minimum cut for the network above has the value

A 18   **B** 20   **C** 22   **D** 23   **E** 31

Question 63 ©VCAA 2010 1NQ8 ●●●

A project has 12 activities. The network below gives the time (in hours) that it takes to complete each activity.

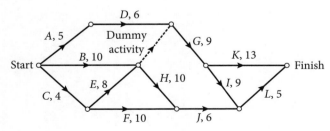

The critical path for this project is

A *ADGK*   **B** *ADGIL*   **C** *BHJL*   **D** *CEGIL*   **E** *CEHJL*

Extended-answer questions

Solutions to this section start on page 287.

64 Brendan's cuts (15 marks)

Question 64.1 (5 marks)

Brendan has heard of the 'minimum cut – maximum flow' theory of directed graphs but doesn't really understand it. To help investigate this idea, he decides to draw a small digraph as follows.

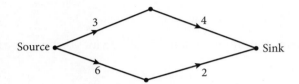

a   Taking the '3–4' path and '6–2' path, separately determine the maximum flow for each section.   1 mark

b   Hence determine the maximum flow from source to sink.   1 mark

c   Draw in all possible cuts and label them A, B, C, etc.   1 mark

d   Determine the capacity of each cut.   1 mark

e   Which cut represents the maximum flow?   1 mark

Question 64.2 (7 marks)

Brendan decides to extend his digraph as shown.

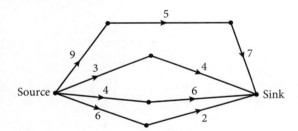

a   Determine the maximum flow from source to sink.   2 marks

9780170465335

Brendan extends his digraph even further to the one shown below. Two cuts, labelled as Cut 1 and Cut 2, have already been made.

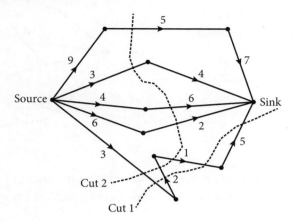

b Explain why Cut 1 is not a valid cut for determining the maximum flow.   1 mark

c Determine the capacity of Cut 2.   1 mark

d Explain why the capacity of Cut 2 is not 22.   1 mark

e Brendan believes that further cuts will be necessary to determine the maximum flow.
Draw in the minimum cut and hence determine the maximum flow from source to sink.   2 marks

Question 64.3 (3 marks)

Brendan's final change to his digraph is to relabel two of the previous weightings as x and y.
Two cuts have been made, labelled as Cut P and Cut Q.

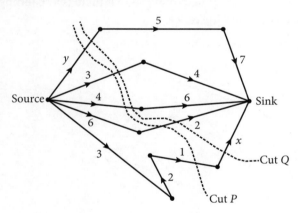

a If Cut Q has a capacity of 15, list all possible values of x and y.   2 marks

b If x is equal to 5 as in Question **64.2**, what would be the greatest value that y could take if
Cut P is now the minimum cut?   1 mark

65 Camping (15 marks)

A large camping ground complex is depicted by the network diagram given. The vertices on the diagram, *A* to *J*, represent amenities blocks and the edges are roads connecting these facilities.

The camping ground manager likes to inspect all the roads each day.

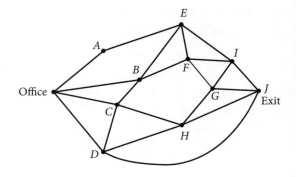

Question 65.1 (3 marks)

a   Could the manager travel over all the roads in the complex, starting and ending at the office, without travelling along any of the roads more than once? Give a reason for your answer.   2 marks

b   What is the name of the type of route that he could take?   1 mark

Question 65.2 (2 marks)

The cleaning team needs to visit each of the amenities' blocks, starting and ending at the office.

a   State a route that they could take so that they do not have to visit each block more than once.   1 mark

b   What is the name of the type of route that they would be taking?   1 mark

Question 65.3 (3 marks)

Water pipes are laid along the roads in the camping ground so that there is a spanning tree connection between the amenities blocks and the office.

a   On the diagram below, draw one of these possible connections.   1 mark

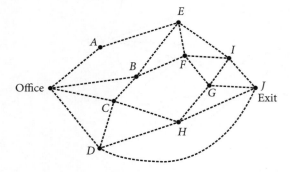

b   If the weightings on the edges below represent distances, in metres, draw on the diagram the minimum spanning tree for the water connections.   2 marks

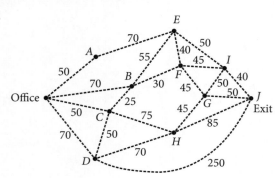

Question 65.4 (2 marks)

If the manager wants to travel from the Office to the Exit at *J*, travelling the shortest possible distance, what route should he take?

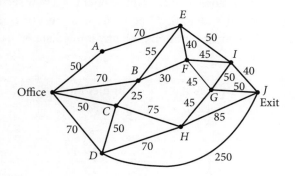

Question 65.5 (3 marks)

At weekends, cars tend to use the roads in the camping ground as a thoroughfare and will travel through the ground from the Office to the Exit at *J*.

Some roads in the camping ground can take more traffic than others and the capacity, in cars per minute, is represented by the weightings on the diagram.

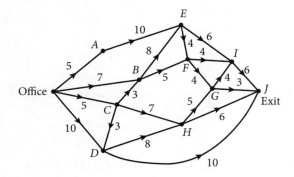

a   What number of cars can enter the camping ground in any minute?   1 mark

b   What is the maximum flow of cars per minute through the camping ground?   2 marks

Question 65.6 (2 marks)

The six workers at the camping ground are multi-skilled and each week they are asked to nominate two of the six tasks that they would prefer to do in that week. The preferences for one week are outlined in the bipartite graph below.

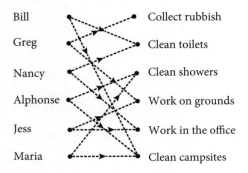

Allocate one task to each worker so that they each get one of their preferences.   2 marks

66 Crashing by design (15 marks)

Question 66.1 (9 marks)

Maccora's beauty salon is undergoing renovations with the aim of opening a new section to cater for facials and waxing. The builder has identified 7 activities that must be completed in order to finish the work. These activities, their durations and any immediate predecessors are given in the following table.

Activity	Duration (weeks)	Immediate predecessors
A	3	–
B	6	A
C	6	A
D	4	A
E	3	B
F	1	C, E
G	2	D, F

a The diagram below shows part of the project network, but is incomplete. Complete the network by including the remaining three activities.   3 marks

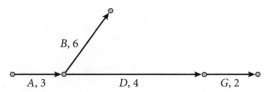

b **i** Determine the critical path for this network.   1 mark

   ii Explain the significance of the critical path in the context of this problem.   1 mark

c The (incomplete) table below shows the Earliest start time (EST) and Latest start time (LST) for each activity. Complete the table by filling in the two missing values.   2 marks

Activity	Earliest start time	Latest start time
A	0	0
B	3	3
C	3	
D	3	9
E	9	9
F		12
G	13	13

d Determine the minimum number of weeks it will take to complete the entire project.   1 mark

The owner of Maccora's beauty salon is unhappy about the length of time for this project and wishes to speed up the process.

e The builder initially offers to reduce the time of Activity C to 4 weeks. Explain why this will not be of any benefit to the salon's owner.   1 mark

Question 66.2 (6 marks)

After extensive negotiations, the builder offers to speed up some of the activities; however, this will result in increased costs to the salon owner. The following table shows which activities may be reduced in time, the cost per week of this reduction and the maximum possible time reduction.

Activity	Cost per week ($)	Maximum reduction (weeks)
B	4000	3
C	3000	2
D	2000	2
E	1000	1

The salon owner studies this table and the original project network in detail. She wants to reduce the length of the project in the most cost-efficient way. She begins to list the activities required to save various numbers of weeks and the costs of these reductions.

Time saved (weeks)	Activities reduced in time	Cost ($)
1	E	1000
2	E, B	5000
3	E, B	
4		

a Explain why the maximum time that can be saved is 4 weeks. 1 mark

b Complete the table above. 3 marks

c Draw the new project network that allows for the maximum reduction of 4 weeks. 1 mark

d Has there been any change to the critical path? 1 mark

67 Landscape (15 marks)

A landscape architect has identified eight major tasks required in the project of designing and constructing a garden. The tasks A to H, their duration and immediate predecessors are shown in the table below.

Activity	Duration (days)	Immediate predecessors
A	7	–
B	10	A
C	8	B
D	2	B
E	3	D
F	4	D
G	3	E, F
H	2	C, G

Question 67.1 (7 marks)

a Two of the activities remain unlabelled on the diagram below. Use the given table to label these activities on the network diagram.   2 marks

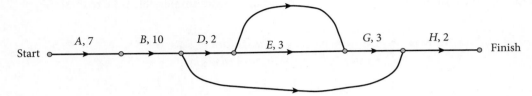

b On the diagram below, complete a forward and backwards scan for each of the activities in the project, placing the correct values in the rectangles provided.   2 marks

Earliest start time	Latest start time

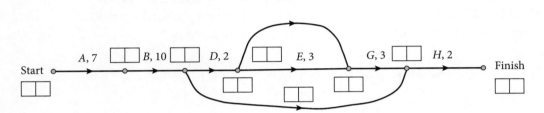

c Write down the critical path for this project.   1 mark

d State the minimum completion time for the project.   1 mark

e State the float time for activity E.   1 mark

Question 67.2 (5 marks)

The project completion time can be reduced at an extra cost. The table below outlines the activities, their possible reduction times, and the cost per day of reducing that activity.

Activity	Possible reduction (days)	Cost of reduction per day
B	3	$400
E	1	$300
G	1	$300

a Find the maximum possible reduction that can be made to the overall finishing time and the cost of this reduction.   3 marks

b If a maximum of $1000 can be spent on reducing the time of completion, which tasks should be reduced and what is the new minimum finishing time?   2 marks

Question 67.3 (3 marks)

The diagram shows the layout for the watering system, with
the eleven nodes representing sprinklers and the weighting
on the edges representing the distance, in metres, between
the sprinklers.

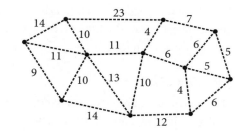

For the watering system to work, each sprinkler needs to be connected to another sprinkler
by a pipe.

a   On the diagram above, indicate the edges that are to be included so that the minimum
    length of pipe is used to connect the sprinklers.   2 marks

b   Find the minimum length of pipe required to connect the sprinklers.   1 mark

68 Neighbours (15 marks)

A rectangular area of land is subdivided into eleven residential blocks of land, as shown
on the diagram below.

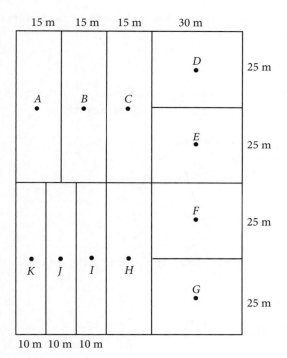

Question 68.1 (4 marks)

a   If, on the diagram above, A to K represent the eleven households and the lines represent
    the boundaries of the blocks, construct a network graph with eleven vertices, A–K,
    where the edges represent 'shares a boundary with'.   2 marks

b   If four streets surround the subdivision, construct a network diagram where the vertices
    are the households A to K and the edges represent 'shares a street boundary with'.   2 marks

Question 68.2 (3 marks)

If a house is considered to be in the centre of its block of land and there is to be one pipeline connecting each of the houses, draw a network diagram showing the connecting pipe that has the minimum length.

Question 68.3 (4 marks)

On a particular subdivision of land, eight rectangular areas of land have been developed as shown in the diagram below. The lines represent the streets between the areas of land.

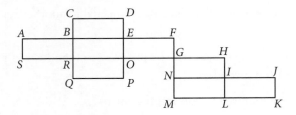

a The postman needs to travel along each of these streets each day to deliver the mail, starting and ending at corner *A*. Explain why he will not be able to do his delivery without retracing his steps along some of the roads.   2 marks

b Which roads will he have to travel along twice so that he delivers the mail while travelling the least distance?   2 marks

Question 68.4 (4 marks)

The subdivision of land is in a district where there are several roads connecting five towns. The roads and towns, labelled *P*, *Q*, *R*, *S* and *T*, are shown on the map below.

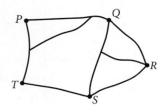

a Complete the network diagram below, where the edges represent the connections between the five towns on the map.   2 marks

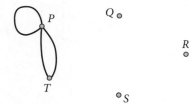

b Construct an adjacency matrix, *M*, for the network diagram above.   2 marks

69 Bridging the gap (15 marks)

In part of a large river there are three islands as shown on the diagram below.

Bridges, as shown, connect the islands and each other to the north and south banks of the river.

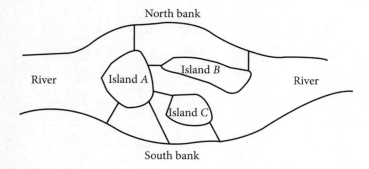

Question 69.1 (9 marks)

a The diagram below contains the vertices of a network diagram where the vertices represent the north and south banks and the islands A, B and C. Complete the diagram by drawing in the edges, where an edge represents a bridge connecting the vertices.   2 marks

(North bank) N

(Island A) A •   • B (Island B)

• C (Island C)

S (South bank)

b State the degree of the vertex A.   1 mark

c Does the graph from part **a** contain an Eulerian circuit? Give a reason for your answer.   2 marks

d Complete the following.

  i An Eulerian circuit starting at the north bank will require an additional bridge joining
    _____ and _____.

  ii This Eulerian circuit is $N \rightarrow B \rightarrow A \rightarrow C \rightarrow S \rightarrow$ _____.   2 marks

e Complete the following.

  i An Eulerian circuit starting at the north bank will require the removal of a bridge joining _____ and _____.

  ii This Eulerian circuit is $N \rightarrow B \rightarrow A \rightarrow C \rightarrow S \rightarrow$ _____.   2 marks

Question 69.2 (2 marks)

The weighted graph shows the distances along roads connecting the banks of the river and the islands.

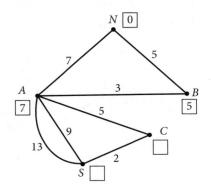

Use Dijkstra's algorithm to complete the shortest path to the vertices C and S from the vertex N.

Question 69.3 (4 marks)

Tolls are charged for travelling on each of the bridges.

The table below gives the charge, in dollars, for travelling one way along a bridge.

		From				
		N	*A*	*B*	*C*	*S*
To	*N*	0	1.80	2.00	0	0
	A	1.80	0	2.40	3.20	1.50
	B	2.00	2.40	0	0	0
	C	0	3.20	0	0	1.50
	S	0	1.50	0	1.50	0

a Complete the graph below by adding the cost weightings to each of the edges.   1 mark

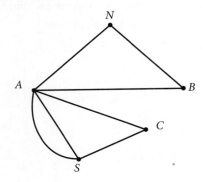

b Use Dijkstra's algorithm to find the minimum cost to each of the vertices when travelling from vertex *N*.   2 marks

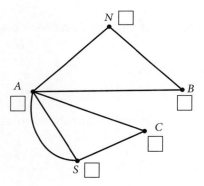

c Determine the minimum toll cost for travelling from *N*, visiting each of the islands *A*, *B* and *C* in any order, and finishing at *S*.   1 mark

70 La Principessa (15 marks)

Question 70.1 (10 marks)

Daniel is a chef at 'La Principessa' Italian restaurant. Each afternoon, he does a short shift to prepare for the evening ahead. The five tasks that are routinely done are listed in the following table, together with their durations. The immediate predecessors are not listed.

Task	Description	Duration (minutes)	Immediate predecessors
A	Peel vegetables	25	
B	Make garlic butter	5	
C	Chop vegetables	20	
D	Prepare garlic bread	10	
E	Arrange ingredients	15	

The project network is shown below.

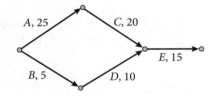

a Complete the column for immediate predecessors in the table.   2 marks

b Which of the activities in this project are critical activities (activities on the critical path)?   1 mark

c What is the minimum time for this project to be completed?   1 mark

d Which of the activities could be delayed without delaying the entire project?   1 mark

e Complete the following table of earliest start times (EST) and latest start times (LST).   3 marks

Activity	EST	LST
A	0	0
B	0	
C	25	25
D		
E	45	45

f Which non-critical activity has the greatest 'float time'? Give its value and interpret this in relation to the problem.   2 marks

Question 70.2 (5 marks)

The four members of the waiting staff at La Principessa are Stephen, Peter, Michelle and Monica. Before work one evening, each will travel from home to a supplier and then on to the restaurant. The following table shows the total distance that each would have to travel in kilometres.

	Supplier 1	Supplier 2	Supplier 3	Supplier 4
Stephen	14	9	20	12
Peter	15	10	23	13
Michelle	9	11	27	9
Monica	15	16	21	12

Using the Hungarian algorithm and a bipartite graph, assign each person to a supplier so as to minimise the overall distance that must be travelled.

71 Security (15 marks)

A security guard is required to patrol along each of the paths in a factory complex. The paths are the edges on the network diagram below and the weightings on the edges represent distance in metres.

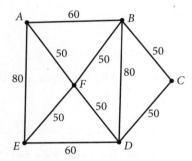

Question 71.1 (7 marks)

a If the guard must start at vertex A, state two routes that would take the guard along each of the paths.   2 marks

b Explain why he will not be able to travel along each of the paths, starting and ending at A, without travelling along one of the paths twice.   2 marks

c Which path will he have to travel along twice so that he walks the minimum distance?   1 mark

d What is the minimum distance that the guard can walk, starting and ending at vertex A?   2 marks

Question 71.2 (4 marks)

At another location, the security guard is required to check each of the entrances, marked *P*, *Q*, *R*, *S*, *T*, *U*, *V* and *W*, on the network diagram below. The weightings on the edges represent distance in metres.

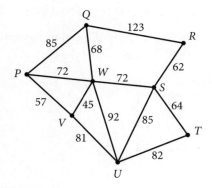

The guard must start and finish at entrance *P*.

a State two routes that he could take so that he does not visit any of the other entrances more than once. 2 marks

b Find the route that he should take so that he walks a minimum distance. 2 marks

Question 71.3 (4 marks)

The security firm has four security guards and four locations to patrol. Security guards are paid for the distance that they travel to their work location, so the security firm is keen to minimise this cost. The table below gives the distance, in kilometres, from the guard's home to each of the four work locations, *A*, *B*, *C* and *D*.

<table>
<thead>
<tr><th rowspan="2"></th><th rowspan="2"></th><th colspan="4">Location</th></tr>
<tr><th>A</th><th>B</th><th>C</th><th>D</th></tr>
</thead>
<tbody>
<tr><td rowspan="4">Guard</td><td>Theo</td><td>8</td><td>12</td><td>18</td><td>13</td></tr>
<tr><td>Reg</td><td>7</td><td>11</td><td>16</td><td>12</td></tr>
<tr><td>Tanya</td><td>12</td><td>11</td><td>17</td><td>15</td></tr>
<tr><td>Jack</td><td>8</td><td>13</td><td>15</td><td>17</td></tr>
</tbody>
</table>

a Using the Hungarian algorithm or otherwise, find the two possible allocations of guards to locations so that the cost will be a minimum for the security firm. 3 marks

b What is the minimum total distance travelled to the four locations by the guards? 1 mark

72 VCAA 2010 Exam 2 (15 marks)

In a competition, members of a team work together to complete a series of challenges.

Question 72.1 (2 marks) ©VCAA 2010 2NQ1

The members of one team are Kristy (K), Lyn (L), Mike (M) and Neil (N).

In one of the challenges, these four team members are only allowed to communicate directly with each other as indicated by the edges of the following network.

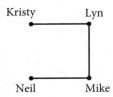

The adjacency matric below also shows the allowed lines of communication.

$$
\begin{array}{cccc}
K & L & M & N
\end{array}
$$
$$
\begin{bmatrix}
0 & 1 & 0 & 0 \\
1 & 0 & 1 & 0 \\
0 & f & 0 & 1 \\
0 & g & 1 & 0
\end{bmatrix}
\begin{array}{c}
K \\
L \\
M \\
N
\end{array}
$$

a Explain the meaning of a **zero** in the adjacency matrix.   1 mark

b Write down the values of f and g in the adjacency matrix.

$f =$ _____

$g =$ _____   1 mark

Question 72.2 (3 marks) ©VCAA 2010 2NQ2

The diagram below shows a network of tracks (represented by edges) between checkpoints (represented by vertices) in a short-distance running course. The numbers on the edges indicate the time, in minutes, a team would take to run along each track.

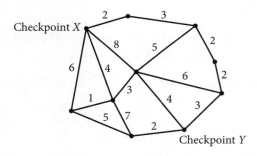

Another challenge requires teams to run from checkpoint X to checkpoint Y using these tracks.

a What would be the shortest possible time for a team to run from checkpoint X to checkpoint Y?   1 mark

b Teams are required to follow a route from checkpoint X to checkpoint Y that passes through each checkpoint once only.

    i What mathematical term is used to describe such a route?   1 mark

    ii On the network diagram below, draw in the route from checkpoint X to checkpoint Y that passes through every checkpoint once only.   1 mark

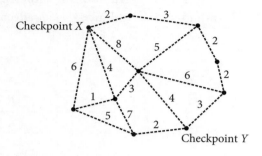

Question 72.3 (4 marks) ©VCAA 2010 2NQ3 ●●●

The following network diagram shows the distances, in kilometres, along the roads that connect six intersections A, B, C, D, E and F.

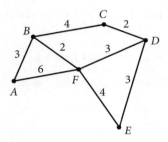

a If a cyclist started at intersection B and cycled along every road in this network once only, at which intersection would she finish?   1 mark

b The next challenge involves cycling along every road in this network at least once. Teams have to start and finish at intersection A.

The blue team does this and cycles the shortest possible total distance.

    i Apart from intersection A, through which intersections does the blue team pass more than once?   1 mark

    ii How many kilometres does the blue team cycle?   1 mark

c The red team does not follow the rules and cycles along a bush path that connects two of the intersections.

This route allows the red team to ride along every road only once.

Which two intersections does the bush path connect?   1 mark

Question 72.4 (6 marks) ©VCAA 2010 2NQ4 ●●●

In the final challenge, each team has to complete a construction project that involves activities *A* to *I*.

Table 1

Activity	EST (minutes)	LST (minutes)	Duration (minutes)	Immediate predecessor
A	0	0	5	–
B	5	5	6	*A*
C	5	6	4	*A*
D	11	11	2	*B*
E	5	9	7	*A*
F		10	6	*C*
G	9	13	1	*C*
H	13	13	3	*D*
I	10	14	2	*G*

Table 1 shows the earliest start time (EST), latest start time (LST) and duration, in minutes, for each activity.

The immediate predecessor is also shown. The earliest start time for activity *F* is missing.

a What is the least number of activities that must be completed before activity *F* can commence?   1 mark

b What is the earliest start time for activity *F*?   1 mark

c Write down all the activities that must be completed before activity *G* can commence.   1 mark

d What is the float time, in minutes, for activity *G*?   1 mark

e What is the shortest time, in minutes, in which this construction project can be completed?   1 mark

f Write down the critical path for this network.   1 mark

73 VCAA 2011 Exam 2 (15 marks)

Question 73.1 (4 marks) ©VCAA 2011 2NQ1 ●●

Aden, Bredon, Carrie, Dunlop, Enwin and Farnham are six towns.

The network shows the road connections and distances between these towns in kilometres.

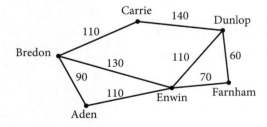

a In kilometres, what is the shortest distance between Farnham and Carrie?   1 mark

b How many different ways are there to travel from Farnham to Carrie without passing through any town more than once?   1 mark

An engineer plans to inspect all of the roads in this network.

He will start at Dunlop and inspect each road only once.

c At which town will the inspection finish? 1 mark

Another engineer decides to start and finish her road inspection at Dunlop.

If an assistant inspects **two** of the roads, this engineer can inspect the remaining six roads and visit each of the other five towns only once.

d How many kilometres of road will the assistant need to inspect? 1 mark

Question 73.2 (2 marks) ©VCAA 2011 2NQ2 ●●●

At the Farnham showgrounds, eleven locations require access to water. These locations are represented by vertices on the network diagram shown below. The dashed lines on the network diagram represent possible water pipe connections between adjacent locations. The numbers on the dashed lines show the minimum length of pipe required to connect these locations in metres.

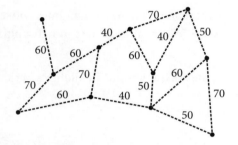

All locations are to be connected using the smallest total length of water pipe possible.

a On the diagram, show where these water pipes will be placed. 1 mark

b Calculate the total length, in metres, of water pipe that is required. 1 mark

Question 73.3 (5 marks) ©VCAA 2011 2NQ3 ●●●

A section of the Farnham showgrounds has flooded due to a broken water pipe. The public will be stopped from entering the flooded area until repairs are made and the area has been cleaned up.

The table below shows the nine activities that need to be completed in order to repair the water pipe. Also shown are some of the durations, Earliest Start Times (EST) and the immediate predecessors for the activities.

Activity	Activity description	Duration (hours)	EST	Immediate predecessor(s)
A	Erect barriers to isolate the flooded area	1	0	–
B	Turn off the water to the showgrounds		0	–
C	Pump water from the flooded area	1	2	A, B
D	Dig a hole to find the broken water pipe	1		C
E	Replace the broken water pipe	2	4	D
F	Fill in the hole	1	6	E
G	Clean up the entire affected area	4	6	E
H	Turn on the water to the showgrounds	1	6	E
I	Take down the barriers	1	10	F, G, H

a What is the duration of activity *B*? 1 mark

b What is the earliest start time (EST) for activity D? 1 mark

c Once the water has been turned off (Activity B), which of the activities C to I could be delayed without affecting the shortest time to complete all activities? 1 mark

It is more complicated to replace the broken water pipe (Activity E) than expected. It will now take four hours to complete instead of two hours.

d Determine the shortest time in which activities A to I can now be completed. 1 mark

Turning on the water to the showgrounds (Activity H) will also take more time than originally expected. It will now take five hours to complete instead of one hour.

e With the increased duration for Activity H and Activity E, determine the shortest time in which activities A to I can be completed. 1 mark

Question 73.4 (4 marks) ©VCAA 2011 2NQ4 ●●●

Stormwater enters a network of pipes at either Dunlop North (Source 1) or Dunlop South (Source 2) and flows into the ocean at either Outlet 1 or Outlet 2.

On the network diagram below, the pipes are represented by straight lines with arrows that indicate the direction of the flow of water. Water cannot flow through a pipe in the opposite direction.

The numbers next to the arrows represent the maximum rate, in kilolitres per minute, at which stormwater can flow through each pipe.

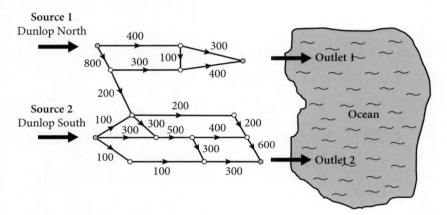

a Complete the following sentence for this network of pipes by writing either the number 1 or 2 in each box.

Stormwater from Source ☐ **cannot** reach Outlet ☐ 1 mark

b Determine the maximum rate, in kilolitres per minute, that water can flow from these pipes into the ocean at

Outlet 1 _____ Outlet 2 _____ 2 marks

A length of pipe, shown in **bold** on the network diagram below, has been damaged and will be replaced with a larger pipe.

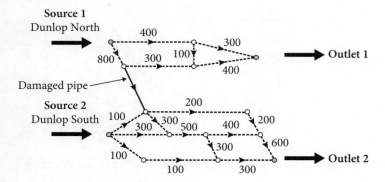

c The new pipe must enable the greatest possible rate of flow of stormwater into the ocean from Outlet 2.

What minimum rate of flow through the pipe, in kilolitres per minute, will achieve this?   1 mark

74 VCAA 2012 Exam 2 (15 marks)

Question 74.1 (5 marks) ©VCAA 2012 2NQ1 ●●

Water will be pumped from a dam to eight locations on a farm.

The pump and the eight locations (including the house) are shown as vertices in the network diagram below.

The numbers on the edges joining the vertices give the shortest distances, in metres, between locations.

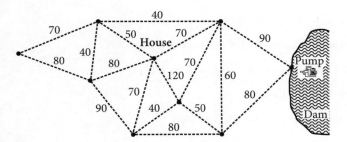

a   **i** Determine the shortest distance between the house and the pump.   1 mark

 ii How many vertices on the network diagram have an odd degree?   1 mark

 iii The total length of all edges in the network is 1180 metres.

  A journey starts and finishes at the house and travels along every edge in the network.

  Determine the shortest distance travelled.   1 mark

The total length of pipe that supplies water from the pump to the eight locations on the farm is a minimum.

This minimum length of pipe is laid along some of the edges in the network.

b **i** On the diagram below, **draw** the minimum length of pipe that is needed to supply water
to all locations on the farm. 1 mark

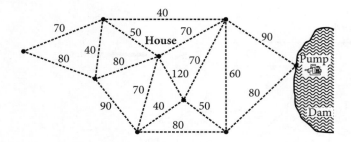

ii What is the mathematical term that is used to describe this minimum length of pipe
in part **i**? 1 mark

Question 74.2 (5 marks) ©VCAA 2012 2NQ2 ⚫⚫⚫

Thirteen activities must be completed before the produce grown on a farm can be harvested.
The directed network below shows these activities and their completion times in days.

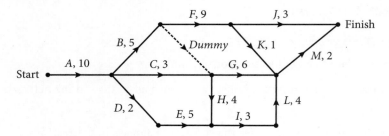

a Determine the earliest starting time, in days, for activity *E*. 1 mark

b A *dummy* activity starts at the end of activity *B*.

Explain why this dummy activity is used on the network diagram. 1 mark

c Determine the earliest starting time, in days, for activity *H*. 1 mark

d In order, list the activities on the critical path. 1 mark

e Determine the latest starting time, in days, for activity *J*. 1 mark

Question 74.3 (5 marks) ©VCAA 2012 2NQ3 ⚫⚫

Four tasks, *W*, *X*, *Y* and *Z*, must be completed.

Four workers, Julia, Ken, Lana and Max, will each do one task.

Table 1 shows the time, in minutes, that each person would take to complete each
of the four tasks.

Table 1

		Worker			
		Julia	**Ken**	**Lana**	**Max**
Task	**W**	26	21	22	25
	X	31	26	21	38
	Y	29	26	20	27
	Z	38	26	26	35

The tasks will be allocated so that the total time of completing the four tasks is a minimum.

The Hungarian method will be used to find the optimal allocation of tasks.

Step 1 of the Hungarian method is to subtract the minimum entry in each row from each element in the row.

Table 2

		Worker			
		Julia	Ken	Lana	Max
Task	W	5	0	1	4
	X	10	5	0	
	Y	9	6	0	7
	Z	12	0	0	9

a Complete step 1 for task *X* by writing down the number missing from the shaded cell in Table 2.

1 mark

The second step of the Hungarian method ensures that all columns have at least one zero.

The numbers that result from this step are shown in Table 3 below.

Table 3

		Worker			
		Julia	Ken	Lana	Max
Task	W	0	0	1	0
	X	5	5	0	13
	Y	4	6	0	3
	Z	7	0	0	5

b Following the Hungarian method, the smallest number of lines that can be drawn to cover the zeros is shown dashed in Table 3.

These dashed lines indicate that an optimal allocation cannot be made yet.

Give a reason why.

1 mark

c Complete the steps of the Hungarian method to produce a table from which the optimal allocation of tasks can be made.

1 mark

Two blank tables have been provided for working if needed.

		Worker			
		Julia	Ken	Lana	Max
Task	W				
	X				
	Y				
	Z				

		Worker			
		Julia	Ken	Lana	Max
Task	W				
	X				
	Y				
	Z				

d Write the name of the task that each person should do for the optimal allocation of tasks.

2 marks

Worker	Task
Julia	
Ken	
Lana	
Max	

75 VCAA 2014 Exam 2 (15 marks)

Question 75.1 (2 marks) ©VCAA 2014 2NQ1

Four members of a train club, Andrew, Brianna, Charlie and Devi, have joined one or more interest groups for electric, steam, diesel or miniature trains.

The edges of the bipartite graph below show the interest groups that these four train club members have joined.

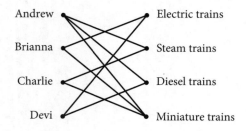

a How many of these four members have joined the steam trains interest group?

1 mark

b Which interest group have both Brianna and Charlie joined?

1 mark

Question 75.2 (4 marks) ©VCAA 2014 2NQ2 ●●

Planning a train club open day involves four tasks.

Table 1 shows the number of hours that each club member would take to complete these tasks.

Table 1

Task	Andrew	Brianna	Charlie	Devi
Publicity	13	12	10	10
Finances	9	10	11	11
Equipment	8	12	11	10
Catering	9	10	11	8

The Hungarian algorithm will be used to allocate the tasks to club members so that the total time taken to complete the tasks is minimised.

The first step of the Hungarian algorithm is to subtract the smallest element in each row of Table 1 from each of the elements in that row.

The result of this step is shown in Table 2.

a Complete Table 2 by filling in the missing numbers for Andrew. 1 mark

Table 2

Task	Andrew	Brianna	Charlie	Devi
Publicity	3	2	0	0
Finances		1	2	2
Equipment		4	3	2
Catering		2	3	0

After completing Table 2, Andrew decided that an allocation of tasks to minimise the total time taken was not yet possible using the Hungarian algorithm.

b Explain why Andrew made this decision. 1 mark

Table 3 shows the final result of all steps of the Hungarian algorithm.

Table 3

Task	Andrew	Brianna	Charlie	Devi
Publicity	4	2	0	1
Finances	0	0	1	2
Equipment	0	3	2	2
Catering	1	1	2	0

c **i** Which task should be allocated to Andrew? 1 mark

  ii How many hours in total are used to plan for the open day? 1 mark

Question 75.3 (4 marks) ©VCAA 2014 2NQ3 ●●

The diagram below shows a network of train lines between five towns: Attard, Bower, Clement, Derrin and Eden.

The numbers indicate the distances, in kilometres, that are travelled by train between connected towns.

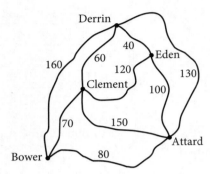

Charlie followed an Eulerian path through this network of train lines.

a   i Write down the names of the towns at the start and at the end of Charlie's path.   1 mark

   ii What distance did he travel?   1 mark

Brianna will follow a Hamiltonian path from Bower to Attard.

b What is the shortest distance that she can travel?   1 mark

The train line between Derrin and Eden will be removed. If one other train line is removed from the network, Andrew would be able to follow an Eulerian circuit through the network of train lines.

c Which other train line should be removed?

In the boxes below, write down the pair of towns that this train line connects.

between ☐ and ☐   1 mark

Question 75.4 (5 marks) ©VCAA 2014 2NQ4

To restore a vintage train, 13 activities need to be completed.

The network below shows these 13 activities and their completion times in hours.

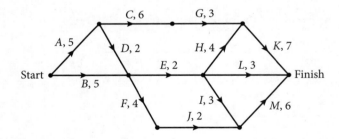

a Determine the earliest starting time of activity *F*.   1 mark

The minimum time in which all 13 activities can be completed is 21 hours.

b What is the latest starting time of activity *L*?   1 mark

c What is the float time of activity *J*?   1 mark

Just before they started restoring the train, the members of the club needed to add another activity, X, to the project.

Activity X will take seven hours to complete.

Activity X has no predecessors, but must be completed before activity G starts.

d What is the latest starting time of activity X if it is not to increase the minimum completion time of the project?   1 mark

Activity A can be crashed by up to four hours at an additional cost of $90 per hour.

This may reduce the minimum completion time for the project, including activity X.

e Determine the least cost of crashing activity A to give the greatest reduction in the minimum completion time of the project.   1 mark

CHAPTER 4 – EXAM PRACTICE

SOLUTIONS

UNIT 3

Chapter 1

Data analysis

Multiple-choice solutions

1 A

For a large group with two distinct age groups, the distribution should be bimodal because the four-year-olds and eight-year-olds will have significantly different modes.

2 D

For options A, B, C and E, the variable can only take distinct values and, therefore, the data is discrete. Volume needs to be measured and therefore is a continuous numerical variable.

3 B

The data tails off to the right and, therefore, is positively skewed.

4 B

The number of weights is found by counting the leaf entries.

5 C

Since the bulk of the distribution is at the lower end, we say that the distribution 'tails off' to the right, indicating a skew to the right or a positive skew.

6 D

Add the number of students that received a mark of 30 or above.

$7 + 11 + 14 + 16 + 18 + 12 = 78$

7 B

Median is the middle value (50% above and below the median).

40% of students received a mark of 40 or less.

46% of students received a mark of 45 or greater.

So, the median is between 40 and 45.

8 C

interquartile range = $50 - 32 = 18$

$1.5 \times IQR = 27$

The boundaries for the outliers are:

$Q_3 + 1.5 \times IQR = 50 + 27 = 77$

$Q_1 - 1.5 \times IQR = 32 - 27 = 5$

Data outside the interval $[5, 77]$ would be outliers.

9 B

There are 18 data values so the median is the average of the ninth and tenth values, i.e. $\frac{16 + 19}{2} = 17.5$

10 B

The median splits the data into two groups of nine so the lower quartile is the fifth value and the upper quartile is the fourteenth value.

interquartile range
= upper quartile – lower quartile
= 14th value – 5th value
= $22 - 10$
= 12

11 C

The upper quartile for this data is 51, so one-quarter of the data has a value greater than or equal to 51. We can conclude, therefore, that more than one-quarter of the data has a value greater than 49.

12 E

In 2000, the lower quartile is $5.50, whereas the median average pay rate in 1980 is $7.50. Not true.

13 B

There are 20 values, therefore the median will be midway between the tenth and eleventh values. From the table, the tenth value is 2 and the eleventh value is 3, so the median will be 2.5.

14 D

If there is only one outlier in the group, then it is either the lowest or the highest score. More information would be needed to say if any of **A**, **B**, **C** or **E** is true.

15 E

Back-to-back stem plots are used to display data between a numerical variable (*speed*) and a two-level categorical variable (*gender of the driver*).

16 C

68% of the students would have a score within one standard deviation of the mean, i.e. in the interval:

$[60 - 6, 60 + 6] = [54, 66]$

55 is in this interval and is closest to the mean.

17 A

The median is 2.

$$\text{mean} = \frac{0 \times 1 + 1 \times 4 + 2 \times 6 + 3 \times 1 + 4 \times 3}{15}$$

$$= \frac{31}{15} = 2.07$$

The mode is 2.

18 C

Since all of the students achieved 2 marks more, the mean must increase by 2. The spread of data, however, will not be altered.

19 C

$\bar{x} - s_x = 8.8 - 2.2 = 6.6$

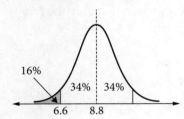

20 D

68% of the eggs will have a mass between 63.8 g and 66.2 g (one standard deviation either side of the mean). 68% of 200 is 136.

21 B

The z-score $= \dfrac{\text{raw score } - \text{ mean}}{\text{standard deviation}}$

$= \dfrac{66.8 - 65}{1.2}$

$= 1.5$

22 C

75 = 64 + 11 = mean + one standard deviation 68% of bell-shaped data can be found within one standard deviation of the mean, i.e. in the interval [53, 75], so half of this can be found in the interval [64, 75].

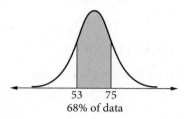

23 D

There are eleven data values that have the value 170 or more (Stems 170, 170*, 180 and 180*).

$\dfrac{11}{25} \times 100 = 44\%$

24 D

The median is the thirteenth value: 167

The lower quartile is the average of the sixth and seventh values:

$\dfrac{164 + 164}{2} = 164$

The upper quartile is the average of the nineteenth and twentieth values:

$\dfrac{173 + 174}{2} = 173.5$

IQR = upper quartile – lower quartile
$= 173.5 - 164$
$= 9.5$

25 C

The standardised score (z-score) is:

$\dfrac{\text{raw score } - \text{ mean}}{\text{standard deviation}} = \dfrac{73 - 60}{6.3} = 2.06$

26 C

Substituting into

$z\text{-score} = \dfrac{\text{raw score } - \text{ mean}}{\text{standard deviation}}$

$-0.7 = \dfrac{\text{raw score} - 64}{10}$

$-0.7 \times 10 = \text{raw score} - 64$

$-7 + 64 = \text{raw score}$

raw data = 57

27 A

$$z = \frac{x - \bar{x}}{s}$$

Substituting gives $-1.3 = \dfrac{x - 23.8}{1.2}$

$$x = -1.3 \times 1.2 + 23.8$$
$$= 22.24$$

28 D

48 = mean − one standard deviation

$\frac{1}{2}$ of 68% = 34% have a weight between 48 kg and 56 kg

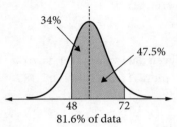

81.6% of data

72 = mean + 2 × standard deviation:

$\frac{1}{2}$ of 95% = 47.5% have a mass between 56 kg and 72 kg

81.5% of 400 = 326

29 C

Use CAS. Data values in L_1.
Frequency in L_2. One variable stats.
$\bar{x}$ and s_x.

30 D

Using CAS:

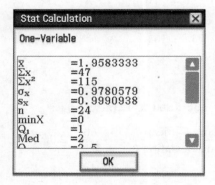

31 E

Use CAS.

32 D

The graph shows moderate positive correlation: y values increase as x values increase. This would correspond to an r value of about 0.6.

33 B

$$\frac{11}{58} = 0.1897$$

18.97% ≈ 19%.

34 B

Both variables have a logical order.

35 C

100 people were questioned, of which 37 are male, therefore:

$$\frac{37}{100} \times \frac{100}{1} = 37\%$$

36 D

63 females were surveyed, of whom 33 intended to vote for the Republicans:

$$\frac{33}{63} \times \frac{100}{1} \approx 52\%.$$

37 E

Parallel boxplots are appropriate whenever displaying the relationship between a numerical variable, *performance on test* and a two-or-more level categorical variable, three classes.

38 B

The points on the scatterplot show moderate negative correlation, which corresponds to an r value of −0.7.

39 C

The results at all five key points are lower on Test *B*. The spread of results for Test *B* is significantly greater, therefore the results are more variable.

40 B

five-number summary:

min 32, Q_1 = 48, median = 64, Q_3 = 76, max = 180

41 D

Half (median) of the Pacific rivers are 100 km in length.

42 C

The graph shows strong positive correlation, which would mean an r value of about 0.8. The coefficient of determination, r^2, would be approximately $0.8^2 = 0.64$.

43 C

Fill in the missing numbers in the table.

Level of mobile phone usage	Age			Total
	12–13	14–15	16–17	
Never	12	14	10	36
Use family phone	37	28	10	75
Phone owner	24	21	40	85
Total	73	63	60	196

Reading from the total column, the number of phone owners is 85.

44 D

Both are categorical variables. There is a logical order to *age* but not to *level of phone usage*.

45 E

37 of the 73 students who are 12 to 13-year-olds use the family phone. As a percentage, this is:

$$\frac{37}{73} \times \frac{100}{1} \approx 50.68\%$$

46 A

$r^2 = 0.65$, so $r = \pm\sqrt{0.65} = \pm 0.81$. An increase in stress is associated with a decrease in productivity, which implies negative correlation, so $r = -0.81$.

47 E

r values fall in the interval $[-1, 1]$; with $+1$ and -1 representing perfect correlation, and 0 representing no correlation.

48 C

Parallel boxplots are used to display data for a numerical variable (*monthly median rainfall*) and two or more level categorical variables (*month of year*).

49 B

Gender is a categorical variable with two categories and *reading ability* (five levels or categories) is also a categorical variable. *Reading ability* is more likely to depend on *gender*, so *gender* is the explanatory variable. A two-way frequency table could be used to display the results of the investigation.

50 B

Strong positive, linear correlation (as indicated by the graph) does not mean that an increase in one variable causes an increase in the other variable. The comment in option **C** is used to describe the coefficient of determination, not the correlation coefficient.

51 D

50% of the students in Class *C* scored more than 65%. In Class *B*, more than 25% scored more than 65%, so doubling this would be more than 50%.

52 B

As x increases, y decreases so there is moderate negative correlation between the variables y and x.

$r \approx -0.7$. Removing (a) would make r closer to -1; a decrease in value. $r^2 = (-0.7)^2 = 0.49$; a value greater than -0.7.

53 A

Assuming that, as the *number of kilometres travelled* increases, the *depth of tread* will decrease, r is negative.

With $r^2 = 0.84$, $r = -0.9$.

Only option **A** shows a correlation near -1.

54 A

There are only two travel times greater than 60 minutes: 61 min (Route *B*) and 71 min (Route *A*).

$$\frac{2}{37} \times 100 \approx 5.41\%$$

55 E

Five-number summaries need to be found for each route. The median for Route *B*'s travel times is 36 min and the median for Route *A*'s travel times is 41.5 min; a difference of 5.5 minutes.

56 D

Option **A** – implies 'cause'.

Option **B** – ± 1 not true.

Option **C** – both 'stray cats' and 'stray dogs' are associated with population increase.

Option **E** – The information does not provide cat owner's behaviour.

57 D

Reading from the graph, the percentage is more than 25%, but not 50%.

58 C

Approximately 75% of people in the 60+ age group are affected by arthritis, compared to about 30% in the 30–<40 age group. 75% is not exactly twice 30%.

59 D

r values are sensitive to outliers, so the r value with the outlier would be negative and approximately −0.6. If the outlier is removed, then the r value would be closer to −1; a decrease in value.

60 C

By direct substitution into the equation:

runs scored = −33.42 + 3.53 × 75

which is closest to 232.

61 B

The line is of the form:

cholesterol level = 0.05 × *age* + *c*

Substituting the values from the point (40, 5.3) gives $c = 3.3$.

62 C

A positive correlation between two variables does not imply that an increase in one variable causes the increase in the other variable.

63 A

Using two points (0, 210) and (10, 100),

$$\text{gradient} = \frac{100 - 210}{10 - 0} = -11$$

average rainfall = 210 − 11 × *temperature range*

64 C

$r = -0.9260$

coefficient of determination = $r^2 = 0.8575$

65 B

two points: (900, 10.7), (1700, 6.7)

$$\text{slope} = \frac{6.7 - 10.7}{1700 - 900}$$

$$= -\frac{4}{800}$$

$$= -0.005$$

66 A

An x^2 transformation will stretch the x scale to linearise the data.

67 B

An x^2 transformation will stretch the x scale to linearise the data. The other options compress the scales.

68 B

This is the definition of the least squares line.

69 C

residual value = actual value − predicted value

The predicted value is found by substituting $Q = 3.5$ in the equation:

$P = 82.5 - 8.6Q$

Predicted P is 52.4; actual P is 46.4, so residual is:

$46.4 - 52.4 = -6$

70 D

Using CAS, key the data into the lists.

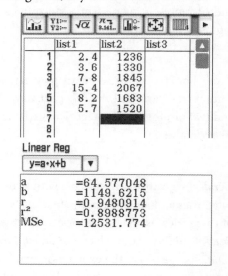

71 D

The explanatory variable is graphed on the horizontal (x) axis. A $\log x$ transformation compresses the x values and this would only linearise graph D.

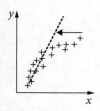

72 A

The residual graph can be seen by dragging up the least squares line until it is horizontal:

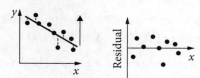

73 B

The slope is negative, -0.62, so can be interpreted as: 'For each increase of one unit in x, y decreases by 0.62 units.'

74 A

The point $(5, 10)$ will almost be an outlier in this set of data. Least-squares lines are sensitive to extreme values and the line will be drawn towards this point. The slope of the new line will increase in value; from -0.62 to about -0.42 and the y-intercept will decrease.

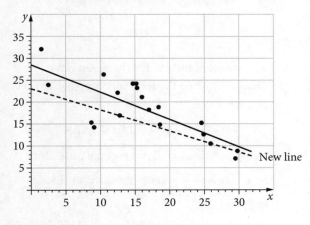

75 C

Either the x scale needs to be compressed ($\log x$) or the y scale needs to be stretched (y^2).

Only alternative given is option **C**.

76 D

Option **B** shows no association, option **C** is linear and option **A** could not be linearised by a log transformation. Log transformations compress the scale to which they are applied. To linearise option **E**, either the x scale or y scale needs to be stretched.

77 B

residual value = actual value − predicted value

predicted value = $0.6 + 2.2 \times 2.4 = 5.88$

actual value = 2

residual = $2 - 5.88 = -3.88$

78 C

Weight is the response variable (y) and height is the explanatory variable. s_x and s_y are the standard deviations of the x (height) and y (weight) values respectively.

The slope of the least-squares regression line is:

$$\frac{rs_y}{s_x} = \frac{0.86 \times 6.8}{5.2}$$
$$= 1.1246$$

79 C

The intercept of the regression line is:

$$\bar{y} - \text{slope} \times \bar{x} = 62 - 1.1246 \times 165.5$$
$$= -124.12$$

where $\bar{x}$ and $\bar{y}$ are the means of the height and weight respectively.

80 B

Irregular fluctuations exist in all time series graphs. Seasonality, one-off events and trends can exist as well as irregular fluctuations.

81 C

The three-point smoothed value for April will be the average of March, April and May.

$$\frac{19 + 22 + 28}{3} = \frac{69}{3}$$
$$= 23$$

82 C

The pattern repeats every four quarters, so it is seasonal. There is an upward trend: as time increases, sales also increase.

83 D

The sum of the seasonal indices for quarterly data will be 4.

$4 - 1.00 - 0.79 - 1.09 = 1.12$

84 D

Month	Jan	Feb	Mar	Apr	May	Jun	Jul	Aug
Number of sedans sold	101	122	137	118	112	104	97	88

Mean of 122, 137, 118 and 112 = 122.25

Mean of 137, 118, 112 and 104 = 117.75

Centring:
$\dfrac{122.25 + 117.75}{2} = 120$

85 D

The time period for Quarter 3 of 2018 will be 11. Substituting this into the equation gives:

de-seasonalised sales $= 44.43 \times 11 + 971$
≈ 1460

86 E

To re-seasonalise the predicted value, we multiply by the seasonal index for that quarter (Quarter 3):

$1460 \times 1.086 = 1585.56$

87 D

A seasonal index of 0.836 indicates that the sales figures for the June quarter are $(1 - 0.836) \times 100 = 16.4\%$ below the average for the year. To de-seasonalise actual figures we divide by the seasonal index.

88 D

Using years 4, 5, 6, 7, 8 the median point is closest to $(6, 4000)$.

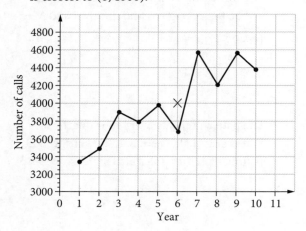

89 B

Month	Jan	Feb	Mar	Apr	May	Jun	Jul	Aug
Number of days	14	12	9	7	4	5	5	8

Moving mean for Feb/Mar is:
$\dfrac{14 + 12 + 9 + 7}{4} = \dfrac{42}{4}$

Moving mean for Mar/April is:
$\dfrac{12 + 9 + 7 + 4}{4} = \dfrac{32}{4}$

The smoothed value, centred, for March is:

$\dfrac{\left(\frac{42}{4} + \frac{32}{4}\right)}{2} = 9.25$

90 E

Three-point moving median for 'time period 4' is the median of 16, 21 and 17, which is 17.

three-point moving means for time period 4 =

$\dfrac{16 + 21 + 17}{3} = \dfrac{54}{3} = 18$

91 A

$0.84 = 1 - 0.16$

This indicates that the number of meals served on Wednesday is 16% less than the daily average.

92 E

$\begin{aligned} \text{de-seasonalised figure} &= \dfrac{\text{actual figure}}{SI} \\ &= \dfrac{108}{0.71} \\ &= 152.1 \end{aligned}$

93 D

The seasonal indices for the seven days will sum to 7 so the total for Saturday and Sunday is

$7 - 0.68 - 0.71 - 0.84 - 1.01 - 1.10 = 2.66$

Using proportions: The seasonal index for Saturday is $\dfrac{190}{190} + 160 \times 2.66 = 1.44$ closest to 1.45.

Alternatively using $\dfrac{\text{long term average}}{\text{seasonal index}}$ are the same for each day.

$\dfrac{89}{0.68} = 131 \qquad \dfrac{93}{0.71} = 131$

So $\dfrac{190}{SI} = 131 \qquad SI = \dfrac{190}{131} = 1.45$

94 E

The seasonal indices will sum to 4, so the missing value is:

$4 - 0.84 - 1.14 - 1.09 = 0.93$

To de-seasonalise, we divide by the seasonal index: $\dfrac{1765}{1.14} = 1548$

95 E

Find the median point using the first three points. Move along one point to find the next median point (using the second, third and fourth points). Continue with this method until you have found the median of the last three points.

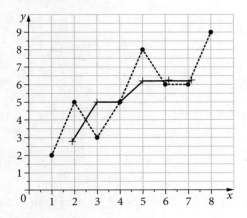

96 C

There is no evidence of seasonality so an increasing trend with irregular fluctuations is shown.

97 A

We divide by the seasonal index to de-seasonalise data. Dividing by 0.96 (a value less than 1) will increase the value, and dividing by 1.11 (a value greater than 1) will decrease the value.

98 D

The yearly average is found by averaging the four figures.

yearly average = 2337.5

Dividing the third quarter figure by the yearly average will give the seasonal index for the third quarter:

$\dfrac{1832}{2337.5} = 0.784$

99 C

Using Q4 2014 to Q3 2015:

$$\text{mean} = \frac{4.5 + 9.6 + 14.5 + 8.6}{4} = 9.3$$

Using Q1 2015 to Q4 2015:

$$\text{mean} = \frac{9.6 + 14.5 + 8.6 + 5.3}{4} = 9.6$$

$$\text{centring} = \frac{9.3 + 9.6}{2} = 9.45$$

100 A

The gap (difference) between the share price of the two companies is increasing over the 20 months.

Extended-answer solutions

101 Approaching normality

Question 101.1

a Half of the total number of students, so 10 000.

b Values of 23 and 37 are one standard deviation on either side of the mean, so 68%.

Values of 16 and 44 are two standard deviations on either side of the mean, so 95%.

numbers of students:

$\dfrac{68}{100} \times 20\,000 = 13\,600$ and $\dfrac{95}{100} \times 20\,000 = 19\,000$

c Recognise that this would be the region which is one standard deviation on either side of the mean (68%) plus another standard deviation on either in the positive or negative direction.

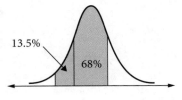

$$\frac{95 - 68}{2} = \frac{27}{2} = 13.5\%$$

Possible ranges: 23 to 44, or 16 to 37.

d $z = \dfrac{x - \bar{x}}{s}$, where x is the raw score, $\bar{x}$ is the mean and s is the standard deviation.

e $z = \dfrac{44 - 30}{7} \Rightarrow z = 2$

Harry is two standard deviations above the mean, i.e. in the top 2.5% of the group.

f Networks: $z = \dfrac{46 - 55}{18} = -0.5$

Matrices: $z = \dfrac{61 - 77}{20} = -0.8$

Despite the higher raw score for Matrices, the z-score shows that Harry is 0.8 standard deviations below the mean, in comparison to 0.5 standard deviations below the mean for Networks. Harry is, therefore, ranked higher in his class on Networks.

Question 101.2

a The mean is 50 because it is halfway between 36 and 64. Since 68% is within one standard deviation from the mean $50 + s = 64$ and $50 - s = 36$ giving:

standard deviation = 14

b A mark of 30 is more likely because it is closer to the mean.

c 100 would have a z-score of

$z = \dfrac{100 - 50}{14} = 3.57$.

While possible, a score well beyond 3 standard deviations from the mean is very unlikely.

102 Card club

Question 102.1

a There appears to be an increasing trend with irregular fluctuations. No seasonal pattern is apparent.

b

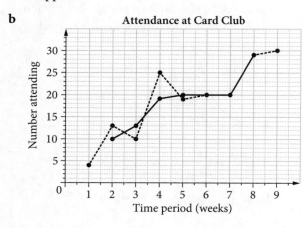

Attendance at Card Club

Question 102.2

a

Time period	1	2	3	4	5	6	7	8	9
Attendance	4	13	10	25	19	20	20	29	30
Smoothed data			14.88	17.63	19.75	21.5	23.38		

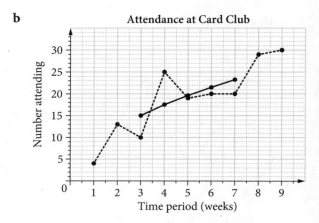

Mean of 25, 19, 20, 20 = 21

Mean of 19, 20, 20, 29 = 22

Centring:

$\dfrac{21 + 22}{2} = 21.5$

b

Attendance at Card Club

c The smoothed graph shows an increasing trend.

Question 102.3

a Entering the data into the lists in CAS, using Calc, and Linear Reg. The Xlist is list1, and the Ylist is list2.

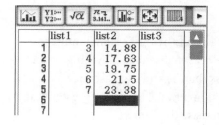

	list 1	list 2	list 3
1	3	14.88	
2	4	17.63	
3	5	19.75	
4	6	21.5	
5	7	23.38	
6			
7			

Linear Reg

$y = a \cdot x + b$

a	=2.087
b	=8.993
r	=0.9957047
r²	=0.9914279
MSe	=0.12553

$attendance = 8.99 + 2.09 \times time\ period$

b For each increase of one week, there is an increase in attendance of two (2.09) people.

c **i** $attendance = 8.99 + 2.09 \times 2 = 13.17 \approx 13$

ii $attendance = 8.99 + 2.09 \times 8 = 25.71 \approx 26$

d

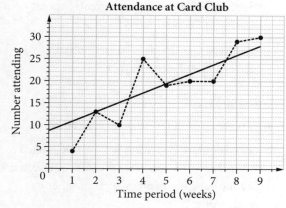

103 Carmelo's gelati

Question 103

a Averaging the four values for 2018 gives a value of 421.

Year	Quarterly average
2016	390
2017	380
2018	421

b Divide each value by its quarterly average.

Year	Quarter 1	Quarter 2	Quarter 3	Quarter 4
2016	1.287	0.879	0.751	1.082
2017	1.282	0.826	0.668	1.224
2018	1.342	0.912	0.713	1.033

c

	Seasonal index
Quarter 1	1.30
Quarter 2	0.87
Quarter 3	0.71
Quarter 4	1.11

d seasonal index = $1 + \dfrac{30}{100} = 1.3$

Sales in Quarter 1 are 30% more than in an average quarter.

e Seasonal indices always average out to 1. If four quarters are used, the indices will add to 4. If twelve months are used, the indices will add to 12.

f Divide each value by the seasonal index for its quarter.

Year	Quarter 1	Quarter 2	Quarter 3	Quarter 4
2016	386	394	413	380
2017	375	361	358	419
2018	435	441	423	392

g Enter data in CAS: 1 to 12 in list1 and the data from the table above in list2. Find the least-squares regression line.

de-seasonalised sales
= 3.24 × *time period* + 377.02

h Substitute 10 for *time period* in the equation.

de-seasonalised sales = 3.24 × 10 + 377.02 ≈ 409

i Residual is the actual value minus the value predicted by the least squares line:

residual = 441 − 409 = 32

j Substitute 16 for *time period* in the equation.

de-seasonalised sales = 3.24 × 16 + 377.02
= 428.86

To forecast actual sales, we must re-seasonalise by multiplying by the seasonal index for Quarter 4:

428.86 × 1.11 ≈ 476

k Extrapolation, because we are predicting outside the original set of values.

104 Fatten-up

Question 104.1

a

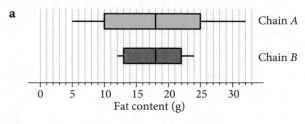

b The distributions of fat content for meals from both fast-food chains are symmetrical and have the same median fat content of 18 g. The fat content for meals from chain *A* is more variable than that for meals from chain *B*; the range for chain *A* is 27 g, compared to 12 g for chain *B*. (The IQR for chain *A* is 15 g compared to 9 g for chain *B*.)

Question 104.2

a There appears to be a strong, positive, linear relationship between the variables *fat content* and *energy*. As fat content increases, the energy increases.

b Pearson's correlation coefficient, *r*, is 0.9446. This supports the observations from the scatterplot having strong, positive correlation between the variables.

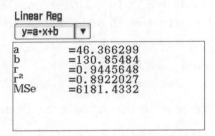

Linear Reg
y=a·x+b ▼

a	=46.366299
b	=130.85484
r	=0.9445648
r²	=0.8922027
MSe	=6181.4332

c $r^2 = 0.8922$, indicating that 89.22% of the variation in energy can be explained by the variation in fat content.

Question 104.3

a *energy* = 130.85 + 46.37 × *fat content*

b energy = 130.85 + 46.37 × 30
　　　　= 1522 kJ (to the nearest kJ)

c Predicted value is given by:

energy = 130.85 + 46.37 × 20
　　　　= 1058 kJ (to the nearest kJ)

actual value = 1001 kilojoules

residual value = actual value − predicted value
　　　　　　　= 1001 − 1058
　　　　　　　= −57 kilojoules

105 Geoff and Helen

Question 105.1

a Using CAS:
mean = 56.5
standard deviation = 13.72

b mean = 63.1
standard deviation = 20.9

c Parallel boxplots are appropriate since we have a numerical variable (*score*) and a two-valued categorical variable (*classes*).

d Shape: Geoff's class is negatively skewed whereas Helen's is approximately symmetric.

Centre: Helen's class has a higher centre (median of 64.5, compared to 57.5).

Spread: Helen's class has a greater spread (range of 78, compared to 58).

Outliers: Geoff's class has an outlier at 90. Helen's class has no outlier.

Question 105.2

a The mean will increase by 2, to 58.5.

b The standard deviation will be unchanged (remains at 13.72). The spread is unaffected by the addition of a constant number to each value.

Question 105.3

a **i** One standard deviation from the mean is the interval:

$[63.1 − 20.9, 63.1 + 20.9] = [42.2, 84]$

11 of 16 marks are within one standard deviation, i.e. 68.75%.

ii 15 of 16 are within [21.3, 104.9]; two standard deviations on either side of the mean, i.e. 93.75%.

iii 16 of 16 are within three standard deviations, i.e. 100%.

b These three percentages are close to 68%, 95% and 99.7%, therefore Helen's class results can be approximated to a normal distribution.

106 Global warming

Question 106

a When we are investigating the association between two numerical variables, we use a scatterplot.

b

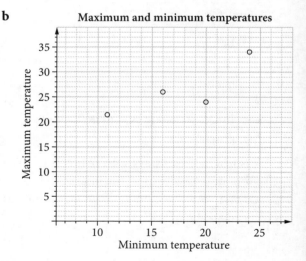

Maximum and minimum temperatures

c There appears to be a strong, positive linear association between the two variables.

d A least squares regression line is appropriate when the data appears linear with no outliers.

e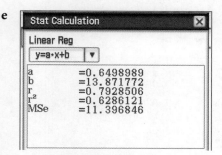

maximum temperature
= 0.65 × *minimum temperature* + 13.87

f Two points that appear to be on the line are (11, 21) and (25, 30).

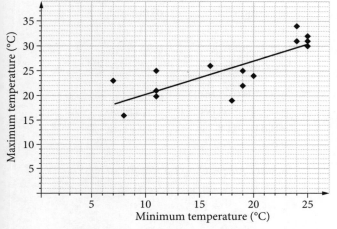

g The two cities that are best predicted by the line are Paris (11, 21) and Bangkok (25, 30).

h Beijing 26.2
London 4
New Delhi 30.1 and 1.9

i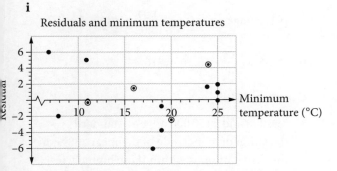

j Random scatter of points suggests the linear model is suitable.

107 One out of the box

Question 107.1

a mean = 64.47; standard deviation = 20.05

b 26, 52, 68, 83, 93

c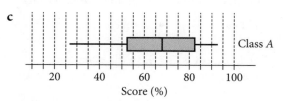

d There is a greater spread in the lower half of results, i.e. the data is negatively skewed and the mean moves towards the extreme values.

e No, all four sections contain 25% of the data regardless of spread.

f IQR = 31

1.5 × IQR = 46.5

fences = 5.5, 129.5

Conclusion: no data outside the interval [5.5, 129.5].

Question 107.2

a The total marks for each class must be the same (1225). 59 is the score which must be added to Class *B* to achieve this.

b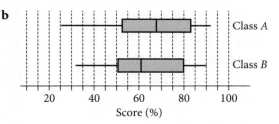

c i Class *A* has a slight negative skew. Class *B* has a slight positive skew.

   ii Class *A* has the higher centre with median 68 compared to 61 for class *B*.

   iii Class *A* has greater spread with higher range and IQR.

d His score would need to be 8% or lower to be an outlier. His score avoided being an outlier by a margin of 24%.

108 Linear behaviour

Question 108.1

a The points on the scatterplot appear to have a moderate, positive linear association.

b About 0.7. Assume linear.

c Can be any line that goes through the points with a positive gradient. A reasonable line would go through $(0, 4)$ and $(14, 15)$.

d The gradient of the line going through $(0, 4)$ and $(14, 15)$ is $\dfrac{15 - 4}{14 - 0} = 0.786 = b$.

The intercept is, $a = 4$, so the equation is $y = 4 + 0.786x$.

e The constant a is the y-intercept: it is the value of y when the value of x is 0.

f The constant b is the increase in y for every unit increase in x.

Question 108.2

a The slope is given by $\dfrac{rs_y}{s_x}$, therefore $\dfrac{0.8 \times 4.5}{5.75}$.

The slope has a value of 0.63 (correct to two decimal places).

b The intercept is given by $\bar{y} - \text{slope} \times \bar{x}$, therefore:

$11.6 - 0.63 \times 9.5 = 5.62$

c The least squares equation is: $y = 0.63x + 5.62$.

Substituting $x = 7.6$, gives:

$y = 0.63 \times 7.6 + 5.62$
$ = 10.41$

d Residual is the actual value minus the value predicted by the regression line:

residual $= 12 - 10.41$
$ = 1.59$

e The prediction must increase by 1.59, therefore the intercept will be:

$5.62 + 1.59 = 7.21$

109 Share and share alike

Question 109.1

a The data needs to be ordered to find the middle value; the median is $0.20.

> **Note**
> CAS can be used to find the median, mean and five-number summary.

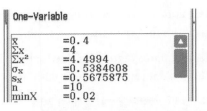

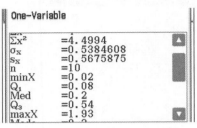

b The mean is $0.40.

c The median is not affected by extreme values, e.g. $1.93 for QR.

d

minimum value	0.02
lower quartile	0.08
median	0.20
upper quartile	0.54
maximum value	1.93

e interquartile range
= upper quartile – lower quartile
= $0.54 - 0.08$
= 0.46

upper fence $= Q_3 + 1.5 \times \text{IQR}$
$ = 0.54 + 1.5 \times 0.46$
$ = 1.23$

Therefore, the value for QR needs to be:

$1.93 - 1.23 = \$0.70$ lower

Question 109.2

a XYZ rose 0.54 to 15.56, therefore was $15.02 at the start of the day.

percentage increase $= \dfrac{0.54}{15.02} \times 100 = 3.6\%$

b $15.56 + \dfrac{5}{100} \times 15.56 = \$16.34$

Question 109.3

a $90 000 and $150 000 are one standard deviation on either side of the mean. 68% of the distribution is within these boundaries.

b $180 000 is two standard deviations above the mean. 95% of values are within two standard deviations from the mean. The remaining 5% is divided equally at the lower end and higher end, therefore 2.5% is above $180 000, leaving 97.5% below.

Question 109.4

68% is one standard deviation on either side of the mean. The remaining 32% is divided equally on either side, i.e. 16% on each side. Therefore, 84% is lower than the value that is one standard deviation above the mean. Since $160\,000 is that value, and it is $20\,000 above the mean, the standard deviation is $20\,000.

110 Tough brake

Question 110.1

a

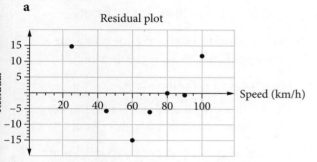

Residual plot

b There is a clear pattern in the residual plot, indicating that the linear model is not a good fit.

Question 110.2

a

Speed (km/h)	(Speed)^2	Stopping distance (m)
25	625	44
45	2025	51
60	3600	62
70	4900	85
80	6400	105
90	8100	118
100	10000	144

b

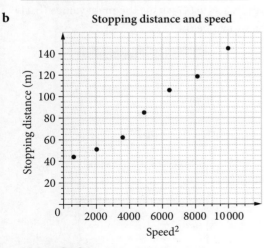

Stopping distance and speed

c Using CAS:

stopping distance = $30.554 + 0.011 \times speed^2$

d stopping distance = $30.554 + 0.011 \times 110^2$

= $163.7\,\text{m}$

Question 110.3

a Predicted stopping distance for a speed of 100 km/h is:

$30.554 + 0.011 \times 100^2 = 140.55$

actual value = $144\,\text{km/h}$

residual value = actual value – predicted value

= $144 - 140.55$

= 3.45

> **Note**
> If the residual feature on the CAS is used, then the value will be 2.61. Both values round to 3.

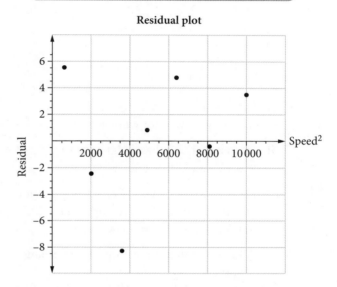

Residual plot

b If the points on the residual plot are randomly scattered about the zero line, as they are in this case, then this indicates that the model is a good fit.

111 Two-way

Question 111.1

a *gender*: It appears that opinion may depend on gender, so *opinion* is the response variable, and *gender* is the explanatory variable.

b Categorical (the data recorded for these variables will be a category, not numerical.)

c

First row (For)	30	29	59
Second row (Against)	28	13	41

d To enable us to see the proportion of males and the proportion of females that voted for or against.

e First column (male): 51.7% for and 48.3% against.

Second column (female): 69.0% for and 31.0% against.

f Yes, opinion on this matter appears to be gender related. 69.0% of women voted for the proposal, compared to 51.7% of men; a difference of 17%.

Question 111.2

a Using stems 1, 2, 3 … 7

> **Note**
> The stem plot can be drawn 'upside-down' compared to the one shown below.

Men	Age	Women
8	1	9
3 1	2	1 2 4 8 9
7 1	3	3 8 9
9 8 5	4	1 6
5 1 1	5	1 7
2 0	6	5
6 2	7	1

b **i** The data is symmetric for men and positively skewed for women.

   ii The data for men has a higher centre. The median is 49 (compared to 38).

   iii The data for men has only a slightly greater spread. The range is 58 (compared to 52), and the IQR is 29 (compared to 27).

   iv Neither distribution has any extreme values or outliers.

112 Vital statistics

Question 112

a

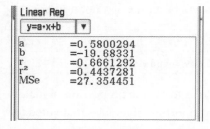

Pearson's correlation coefficient: $r = 0.6661$

b A value for r of 0.6661 represents a moderately strong linear relationship between height and weight.

c Coefficient of determination, r^2, is 0.4437.

d 44.37% of the variation in weight can be explained by the variation in height.

e Lowest value for r^2 is 0.

Highest value for r^2 is 1.

f *weight* = −19.68 + 0.58 × *height*
(see screencap from part **a**)

g weight = −19.68 + 0.58 × 202 = 97.48 kg

h The residual is the actual value minus the value predicted by the regression line:

residual = 95 − 97.48 = −2.48

i Enter the regression formula in list3 as shown below, then calculate residuals in list4.

The missing values of predicted weights are in list3 in the first screen cap.

The missing residuals are in list4 in the second screen cap.

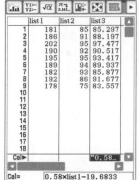
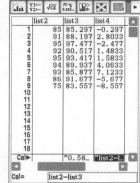

j The residual plot is a scatterplot of the residuals (list4) versus the mass values (list1) as shown below.

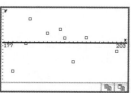

k The random scatter of the points indicates the linear model is appropriate.

113 VCAA 2009 Exam 2

Question 113.1

a

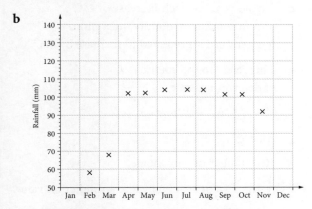

8 9 10 11 12 13 14 15 16 17 18 19 20
Number of rainy days

b **i** There are 12 months so the median is the average of the 6th and 7th dot plot.

So $\dfrac{15 + 16}{2} = 15.5$

ii All but one month, so

$\dfrac{11}{12} \times 100$

$= 92\%$

Question 113.2

a November

b

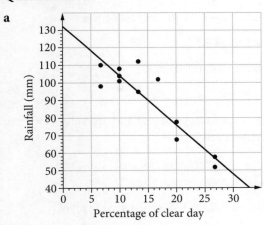

c Rainfall was lower in the first 3 months then consistent from April onwards.

Question 113.3

a

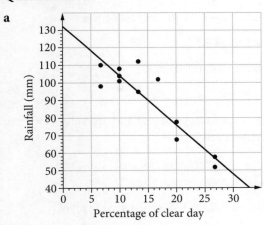

b rainfall = $131 - 2.68 \times 35$

$= 37.2$

c **i** 80.81% of the variation in *rainfall* can be explained by the variation in *percentage of clear days*.

ii If $r^2 = 0.8081$ then $r = \pm\sqrt{0.8081}$

$= -0.899$

Correlation is negative.

Question 113.4

a The seasonal indices will sum to 4. So the *SI* for spring will be:

$4 - 0.78 - 1.05 - 1.07 = 1.1$

b de-seasonalised value $= \dfrac{188}{0.78}$

$= 241$ mm

c $1.05 = 1 + \dfrac{5}{100}$.

So rainfall in autumn is 5% above average.

114 VCAA 2010 Exam 2

Question 114.1

a Australia has 24% women ministers.

11 countries have a percentage > 24%

$\dfrac{11}{22} = 0.5$

b median = 28 (the data is ordered so the median is the average of 24 and 32)

range = 56 (the largest minus the smallest is 56 – 0)

interquartile range = 17 (the 6th value – the 17th value = 38 – 21 = 17)

c Canada has a percentage of 16.

Stem 10(s)	Leaf (units)
0	0
1	2 4 ⑥
2	0 1 3 3 4 4 4
3	2 3 3 6 7 8
4	3 4 7 8
5	6

d As the data distribution is symmetric with no outliers.

Question 114.2

a *Male income* is the explanatory (independent) variable.

b This refers to the slope of the east-squares line.

$0.35 \times \$1000$

$= \$350$

c **i** $13\,000 + 0.35 \times 15\,000$

$= \$18\,250$

ii As we are predicting outside the data range i.e. extrapolation.

Question 114.3

a

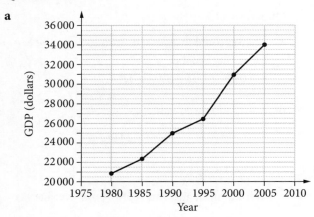

b increasing trend

c $GDP = 20\,000 + 524 \times time$

d Use CAS and 2007 as time period 27. Substituting in the equation:

GDP = $20\,000 + 524 \times 27$
= $34\,148$

error = $34\,900 - 34\,148$
= 752

115 VCAA 2012 Exam 2

Question 115.1

a **i** The median is the average of the 15th and 16th values, which are both 20°C. So 20°C is the median.

ii Seven of the days have maximum temperatures less than 16.

So $\frac{7}{30} \times 100 = 23.3\%$

b $14 = 9.5 + 2 \times 2.25$
= mean + 2 × standard deviation

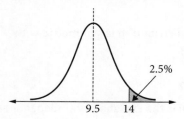

So 97.5% of days have temperatures less than 14°C.

Question 115.2

a

Maximum temperature (°C) vs Minimum temperature (°C) scatter plot

b The vertical intercept is (0, 13). When the minimum temperature is 0°C, the maximum temperature is predicted to be 13°C.

c moderate, positive

d On average, for each increase of 1°C in minimum temperature, maximum temperature is predicted to increase by 0.67°C.

e This refers to the coefficient of determination, r^2.

$r^2 = (0.630)^2 = 0.3969 \approx 0.4$

Answer is 40%.

40% to nearest percentage

f residual value = actual value – prediction value

predicted max. temp = $13 + 0.67 \times 11.1$
= 20.44

residual = $12.2 - 20.4$
= -8.2
= $-8°C$ to nearest degree

Question 115.3

a south-east, north-east

b 8 values: 2, …, …, 3, 4, …, …, 4

$Q_1 = 2$ = average of 2nd and 3rd values so these values must be 2.

$Q_3 = 4$ = average of 6th and 7th values so these values must be 4.

Hence 2, 2, 2, 3, 4, 4, 4, 4 are the eight wind speeds.

Question 115.4

a Using CAS

$(ws\,3.00\,pm)^2 = 3.4 + 6.6 \times ws\,9.00\,am$

b $(ws\ 3.00\,pm)^2 = 3.4 + 6.6 \times 24$

$\qquad\qquad\qquad = 161.8$

$ws\ 3.00\,pm = \sqrt{161.8} = 12.7$

13 km/h to nearest whole number

116 VCAA 2014 Exam 2

Question 116.1

a 19%; reading from the graph

b 29 440 000; 23% of 128 000 000

c All 3 countries have approximately equal percentages. Australia has 67% and Japan and India both have 64%.

Question 116.2

a The response (dependent) variable is *population*.

b

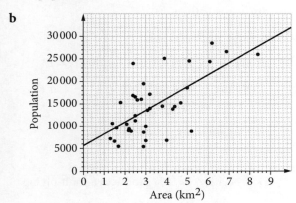

c On average, for each increase of 1 km² in area, population is predicted to increase by 2680.

d **i** Using the least squares line to predict the population in Wiston:

residual value = actual value – predicted value

predicted value = 5330 + 2680 × 4 = 16 050

residual value = 6690 – 16 050

$\qquad\qquad\qquad = -9360$

ii Refers to the coefficient of determination, r^2

$r = 0668,\ r^2 = 0.4462$

hence 44.6% to one decimal place.

Question 116.3

a 7.7, 7.7; using CAS.

b *population* (1000's) $= 7.7 + 7.7 \times \log_{10}(area)$

$\qquad\qquad\qquad = 7.7 + 7.7 \times \log_{10}(90)$

$\qquad\qquad\qquad = 22.75$

23 000 to nearest thousand

Question 116.4

a weak, negative, linear

b **i** $z = x - \dfrac{\mu}{S_x} = \dfrac{3082 - 4370}{1560} = -0.826$

Answer: –0.8 to one decimal place

ii 2.5% expected to have an area at least two standard deviations above the mean:

2.5% of 38 $= \dfrac{2.5}{100} \times 38 = 0.95$

Answer: 1 to nearest whole number

iii $\bar{x} + 2 \times s_x = 3.4 + 2 \times 1.6 = 6.6$

Reading from the scatterplot there are 2 of the 38 that have an area greater than 6.6 km²

Answer: 2 suburbs

117 VCAA 2015 Exam 2

Question 117.1

a The interval with the highest frequency is 70–75.

b Reading from the graph: 5 + 9 = 14 countries

c 30 out of 183 countries; $\dfrac{30}{183} \times 100 = 16.4\%$, to one decimal place

Question 117.2

a negatively skewed

b As the years increase from 1953 → 1973 → 1993, median increases (approximately) from 51 → 62 → 69.

Question 117.3

a On average, for each increase of 1 year in female life expectancy, male life expectancy is predicted to increase by 0.88 years.

b *male* = 3.6 + 0.88 × 35 = 34.4 years

c 95% of the variation in male life expectancy can be explained by the variation in female life expectancy.

Question 117.4

a strong, positive, linear

b Using CAS: *male* = 9.69 + 0.81 × *female*

Question 117.5

a From 60 to 82, so 22 years.

b **i** Predicted values for 2030

Aust. −451.7 + 0.2657 × 2030 = 87.67

UK −350.4 + 0.2143 × 2030 = 84.63

Answer: 3 years to nearest year

ii As the predictions involve extrapolation i.e. predicting outside the data set.

Chapter 2

Recursion and financial modelling

Multiple-choice solutions

1 D

$$t_2 = 1.5t_1 + 2 = 1.5 \times 2 + 2 = 5$$

$$t_3 = 1.5t_2 + 2 = 1.5 \times 5 + 2 = 9.5$$

2 D

We are looking at a multiplying factor to obtain V_2 from V_1. Hence $\frac{V_2}{V_1} = \frac{135}{120} = 1.125$.

3 C

$$t_3 = 2t_2 - 5 = 11$$

$$16 = 2t_2; t_2 = 8$$

$$t_2 = 2t_1 - 5 = 8$$

$$13 = 2t_1; t_1 = 6.5$$

4 C

With flat rate depreciation, the same amount is depreciated each year, which is represented by graph C.

5 D

The asset has depreciated by

4850 − 1500 = 3350 over 5 years,

so $\frac{3350}{5}$ = $670 per year.

As a percentage of the original value, this is $\frac{670}{4850} \times 100 = 13.8\%$.

6 D

$$\frac{t_2}{t_1} = \frac{1650}{1500} = 1.1; \frac{t_3}{t_2} = \frac{1815}{1650} = 1.1 \text{ so there}$$

is a common ratio of 1.1, which represents a compound interest increase of 10%.

7 B

A reducing balance depreciation of 15% is associated with a multiplying factor of 1 − 0.15 = 0.85.

Each year, the value would be 0.85 times the value the previous year. So $V_4 = 0.85V_3$

8 C

Allowance has increased for 4 years. 2% increase: a multiplying factor of 1.02 Allowance in 2016 is $232 \times 1.02^4 = \$251.12$.

9 B

Since $t_1 = 5$, then

$$t_2 = a \times 5 - 2 = 5a - 2$$

From the sequence, we know that $t_2 = 13$.

Therefore, $13 = 5a - 2$

$$15 = 5a$$

Solving for a, we get $a = 3$.

10 C

Depreciation of 15% produces a multiplying factor of $1 - \frac{15}{100} = 0.85$.

At the beginning of 2015, the value will be 0.85 of the original.

At the beginning of 2016, the value will be $0.85^2 = 0.7225$ of the original.

At the beginning of 2017, the value will be $0.85^3 = 0.6141\ldots$ of the original.

At the beginning of 2019, the value will be $0.85^5 = 0.4437\ldots$ of the original.

11 D

The value will be decreasing each year, not at a constant rate. The decrease each year will be getting smaller. For example:

1st year: 100

2nd year: 100 × 0.84 = 84; a difference of 16

3rd year: $100 \times 0.84^2 = 70.56$;

a difference of 13.44

The decrease is decreasing.

12 C

Multiplying factor of 0.85.

Solve $P \times (0.85)^5 = 4880$

$$P = \frac{4880}{(0.85)^5}$$

$$= \$10\,998.29$$

13 C

Van has depreciated by $32\,000 - \$5000$

$= \$27\,000$ for 200 000 kilometres travel.

$$\frac{\$27\,000}{200\,000\,\text{km}} = 0.135 \text{ dollars per kilometre}$$

$$= 13.5 \text{ cents per kilometre}$$

14 E

Loss in value $= \$16\,450 - \$9750 = \$6700$

2.4 cents $= \$0.024$

$$\frac{6700}{0.024} = 279166 \approx 280000$$

15 C

Decrease of 15%: multiplying factor of

$$1 - \frac{15}{100} = 0.85$$

15% reduction over four years: multiply by 0.85^4. Value of the equipment is $20\,000 \times 0.85^4$.

16 D

The computer is depreciated by the same amount each year.

16% of $4000 = \$640$

So the value is always $640 less than the value of the previous year.

17 D

$$3000 \times R^2 = 1920$$

$$R^2 = \frac{1920}{3000}$$

$$= 0.64$$

$$R = 0.8$$

$$= 1 - 0.20$$

$$= 1 - \frac{20}{100}$$

This represents a depreciation rate of 20%.

18 E

The increase for each time period is the same for simple interest, so the graph will be a series of discrete points in a straight line increasing at a constant rate, i.e., a positive gradient.

19 C

$R = 1 + \dfrac{r}{100}$, where r is the quarterly interest rate.

$$r = \frac{6}{4} = 1.5$$

$$R = 1 + \frac{1.5}{100} = 1.015$$

20 C

$$\text{interest per month} = \frac{7.5}{12} = 0.625$$

$$\text{substituting in interest} = \frac{PrT}{100}$$

$$\text{interest} = \frac{6500 \times 0.625 \times 15}{100} = \$609.375$$

21 A

$$\text{quarterly interest rate} = \frac{6.4}{4} = 1.6\%$$

3 years = 12 quarters

$$R = 1 + \frac{r}{100} = 1 + \frac{1.6}{100} = 1.016$$

using $A = PR^n$: $A = 12\,000 \times 1.016^{12}$

22 A

$$\text{quarterly (3 months) interest rate} = \frac{5.5}{4}$$

$$= 1.375$$

$$= r$$

Using $I = \dfrac{PrT}{100}$

$$\text{interest} = \frac{5000 \times 1.375 \times 1}{100} = 68.75$$

total $= \$5068.75$

23 D

$$r = \frac{7.75}{365} = 0.02123\ldots$$

using $I = \frac{PrT}{100}$

Using CAS to calculate interest:

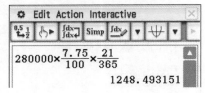

24 D

If $V_0 = 5000$ is the initial amount invested, then V_1 is the amount after 1 year.

$$\begin{aligned} V_1 &= V_0 + 5.2\% \text{ of } V_0 \\ &= V_0 + \frac{5.2}{100} \times V_0 \\ &= V_0\left(1 + \frac{5.2}{100}\right) \\ &= 1.052V_0 \end{aligned}$$

Similarly, $V_2 = 1.052V_1$ and therefore $V_{n+1} = 1.052V_n$.

25 D

The amount of interest is increasing each quarter (eliminates options **B** and **C**), but not at a constant rate (eliminates option **A**). The increase in interest is getting larger each quarter (the increase is decreasing for option **E**).

26 E

$R = 1 + \frac{6}{100} = 1.06$; principal P;

need to find n.

Solve $2P = PR^n$

$2P = P \times 1.06^n$: divide both sides by P;

$2 = 1.06^n$

Use CAS to solve:

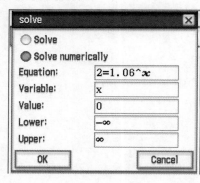

$n = 12$

27 B

annual interest rate = 3%

$R = 1 + \frac{r}{100} = 1 + \frac{3}{100} = 1.03$

For 5 years, $n = 5$.

Use $A = PR^n$: $A = P(1.03)^5$

28 D

using $I = \frac{PrT}{100}$

interest $= \frac{5400 \times 10.5 \times 5}{100} = 2835$

repayment $= \frac{5400 + 2835}{60} = \$137.25$

29 E

$r = \frac{4.95}{12} = 0.4125$

$R = 1 + \frac{0.4125}{100} = 1.004125$

Number of months = 36

Amount accrued $= 15000 \times 1.004125^{36}$
$= 17396.08$

Interest earned $= \$17396.08 - \$15000$
$= \$2396.08$

30 D

Use the Finance solver on CAS:

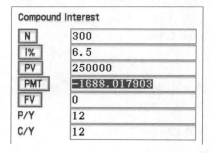

31 C

Use the Finance solver on CAS:

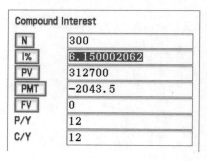

32 E

The monthly interest rate multiplying factor is given and you require the yearly interest rate; $\frac{5.25}{365} = 0.4375$; $\frac{63}{12} = 0.525$ are the two options so yearly interest rate is 6.3%, giving a monthly multiplying factor of $1 + \frac{0.525}{100} = 1.00525$. The repayment amount will be $3022 as $456\,000$ is the first term.

33 C

Use the Finance solver on CAS:

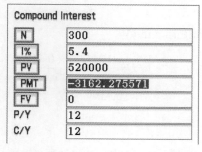

Compound Interest	
N	300
I%	5.4
PV	520000
PMT	-3162.275571
FV	0
P/Y	12
C/Y	12

Compound Interest	
N	100
I%	5.4
PV	520000
PMT	-9506.941105
FV	0
P/Y	4
C/Y	4

With monthly payments, the amount of interest paid is $3162.28 \times 300 - 520\,000 = \$428\,684$.

With quarterly payments, the amount of interest paid = $9506.94 \times 100 - 520\,000 = \$430\,694$.

difference = $\$430\,694 - \$428\,684 = \$2010$

34 D

If the number of payments per year decreases, then more interest is paid over the term of the loan.

If the interest rate does not change, a reduction in the frequency of repayments must result in more interest being charged overall. In effect, this means a longer term of loan or greater repayments to be made.

35 C

The subtraction of 2096 in the recursion relation means that this could be an annuity or a reducing balance loan.

Options **A** and **B** can be eliminated because if the interest rate was 5.5% per year, then the multiplying factor would be 1.055 for yearly compounding and $1 + \frac{5.5}{12 \times 100} = 1.00458$ for monthly compounding.

The multiplying factor for monthly compounding with 6.6% p.a. is $1 + \frac{6.6}{12 \times 100} = 1.055$, as in option **C**.

36 D

Using the Finance solver on CAS, it can be seen that after one payment the amount that Bryan owes is now more than $210\,000$. $615 is less than the interest that accumulates over one month. The graph of the amount still owing will be an increasing graph.

Compound Interest	
N	1
I%	6.65
PV	210000
PMT	-615
FV	-210548.75
P/Y	12
C/Y	12

37 A

The balance will decrease with each yearly payment, so the interest paid will also decrease with each year. This interest paid will not be the same each year, but will be less than the interest paid in the previous year.

38 D

Using the Finance solver on CAS, the extra amount is: $14\,317.88 - $13\,623.37 = $694.51

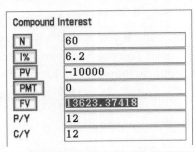

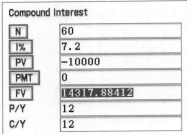

39 E

Monthly interest rate $= \dfrac{7.2}{12} = 0.6\%$

$R = 1 + \dfrac{0.6}{100} = 1.006$

Amount owed after one month
$= AR$
$= 18\,000 \times 1.006$
$= 18\,108$

Subtract repayment of $600.

Amount owing $= $17\,508$

40 B

$P_2 = 1.057P_1$; $P_2 = 7205$, substitute this value in the equation and solve for P_1.

$7205 = 1.057P_1 - 1250$

$8455 = 1.057P_1$: divide both sides by 1.057.

$P_1 = 7999.05$

$P_1 = 1.057P_0 - 1250$

$7999.05 = 1.057P_0 - 1250$

Solve to give $P_0 \approx 8750$.

41 A

4.9% p.a. $= \dfrac{4.9}{12} = 0.408\,333\ldots\%$ per month

Interest on perpetuity $= \dfrac{0.408\,333\ldots}{100} \times 560\,000$
$= 2286.67 per month

42 E

Monthly interest: $r = \dfrac{6.75}{12}$
$= 0.5625$

Substituting in the perpetuity formula:

$P = \dfrac{100Q}{r}$:

$P = \dfrac{100 \times 2200}{0.5625}$
$= $391\,111.11$

43 A

monthly interest $= \dfrac{7.35}{12} = 0.6125$

Substituting in $Q = \dfrac{Pr}{100}$:
$Q = \dfrac{174\,000 \times 0.6125}{100}$
$= 1065.75

44 C

Use the Finance solver on CAS:

Compound Interest	
N	24
I%	5.2
PV	−2000
PMT	−300
FV	9789.168164
P/Y	12
C/Y	12

45 A

Use the Finance solver on CAS:

Compound Interest	
N	36
I%	5.5
PV	−12000
PMT	−662.1519172
FV	40000
P/Y	12
C/Y	12

46 D

$n = 26 \times 2 = 52$

Use the Finance solver on CAS:

Compound Interest	
N	52
I%	5.4
PV	−7500
PMT	−150
FV	16582.19316
P/Y	26
C/Y	26

47 B

Monthly interest rate $= \dfrac{5.4}{12} = 0.45$

$R = 1 + \dfrac{0.45}{100} = 1.0045$

Solving $10\,000 = 8500 \times 1.0045^n$ using CAS gives $n = 37$.

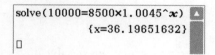

solve(10000=8500×1.0045^**x**)

{x=36.19651632}

48 E

The figure 1.003 75 in the recurrence relation indicates that the amount is increasing each time period as the figure is greater than 1 and then the amount 800 is added. If the 800 was subtracted, it could be a reducing balance loan or an annuity. In this case, the only other option is an investment with additional payments of $800 per month.

49 A

Use the Finance solver on CAS:

Compound Interest	
N	36
I%	5.55
PV	−18000
PMT	−1503.544458
FV	80000
P/Y	12
C/Y	12

Liam will need to deposit approximately $1504 each month to have at least $80\,000 at the end of three years.

50 D

4.5% p.a. = 0.375% interest per month. The investment is going to increase by 0.375% each month; a multiplying factor of 1.003 75, and then $3800 is removed from the account, resulting in the negative sign.

51 B

Use the Finace solver on CAS:

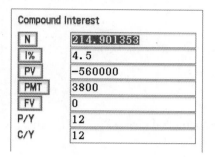

Compound Interest	
N	214.901353
I%	4.5
PV	−560000
PMT	3800
FV	0
P/Y	12
C/Y	12

Helene's investment will last for approximately 215 months = 17.92 years $\approx$ 18 years.

52 C

Multiplying factor of 0.65 is less than 1 and no addition or subtraction.

53 D

$$r_{\text{effective}} = \left[\left(1 + \frac{6.75}{100 \times 365}\right)^{365} - 1\right] \times 100\%$$
$$= 6.98\%$$

54 C

$400\,000 - 34\,572.00 = \$36\,5428.00$

55 C

$47\,372.00 - 8157.81 = \$39\,214.19$

56 C

3.2% of $133\,484.14 = \$4271.49$

57 A

The multiplication factor is $1 - \dfrac{8}{100} = 0.92$.

58 B

$1 + \dfrac{9.6}{1200} = 1.008$.

59 D

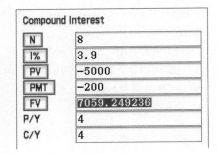

60 E

$250\,000 \times R^6 = 100\,000$

$R \approx 0.86$

61 A

Reducing balance $= P_{n+1} = RP_n$ where $R < 1$

62 C

First, find the interest rate.

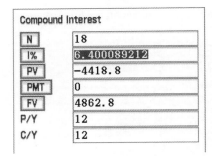

Then find the PV needed to give the FV of $4418.80 six months later.

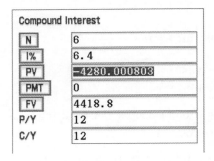

63 E

By recursive calculations or table of values on CAS.

64 D

Interest rate $= \dfrac{720}{180\,000} = 0.4\%$

Amount of interest with repayment 2 is 0.4% of 179 870 = $719.48

Extended-answer solutions

65 Changing houses

Question 65.1

a $\dfrac{6.3}{12} = 0.525\%$ per month

b

Month	Payment ($)	Interest ($)	Principal reduction ($)	Balance ($)
0	0.00	0.00	0.00	235 000.00
1	1577.49	1233.75	323.74	234 676.26
2	1577.49	1232.05	325.44	234 350.82
3	1577.49	1230.34	327.15	234 023.67

c The constant a in the relation will be the multiplying factor associated with the monthly interest.

$$a = 1 + \frac{0.525}{100} = 1.00525$$

The constant b is the amount of the repayment each month, so it will be a negative amount.

$$b = -1557.49$$

d $232 250 to the nearest ten dollars.

The amount of interest over the term of the loan can be found using:

Amount of repayment × number of repayments – amount borrowed

$$= \$1557.49 \times 300 - \$235\,000 = \$232\,247$$

Amortization	
PM1	1
PM2	300
I%	6.3
PV	235000
PMT	−1557.49
P/Y	12
C/Y	12
BAL	2.986513747
INT	
PRN	
ΣINT	−232249.9865
ΣPRN	

e Half the original principal is $117 500, so using this value on CAS and solving for the number of time periods (N) gives 204 months.

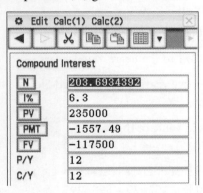

Compound Interest	
N	203.6934392
I%	6.3
PV	235000
PMT	−1557.49
FV	−117500
P/Y	12
C/Y	12

f   i After one repayment, the amount owing is $234 676.26. Subtracting this from the original loan amount:

$$\$235\,000 - \$234\,676.26 = \$323.74$$
$$\approx \$324$$

is the amount of the first repayment that will go towards repaying the loan (the remainder is interest).

ii To calculate this amount, we can find the amount owing after 99 payments and subtract from this the amount owing after 100 payments.

$$\$193\,109.74 - \$192\,566.08 = \$543.66$$
$$\approx \$544$$

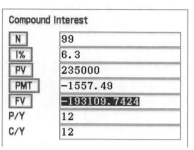

Compound Interest	
N	99
I%	6.3
PV	235000
PMT	−1557.49
FV	−193109.7424
P/Y	12
C/Y	12

Compound Interest	
N	100
I%	6.3
PV	235000
PMT	−1557.49
FV	−192566.0786
P/Y	12
C/Y	12

Question 65.2

a

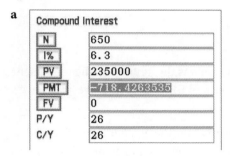

Compound Interest	
N	650
I%	6.3
PV	235000
PMT	−718.4263535
FV	0
P/Y	26
C/Y	26

Using the Finance solver: Vinh will need to repay $718.43 per fortnight.

b $718.43 × 650 − $235 000 = $231 979.50

Making fortnightly repayments, Vinh will pay $231 979.50 in interest compared to $232 247 interest when making monthly repayments. A saving of $267.50.

$270 to the nearest $10.

66 Dave's dilemma

Question 66.1

a 6% of $100 000 = $6000

b 10 × 6000 = $60 000 interest plus $100 000 principal = $160 000

c No effect on total repayments since a flat rate of interest is charged.

Question 66.2

a Using the simple interest formula:

$$I = \frac{\mathrm{Pr}\,T}{100}; \quad 40\,000 = \frac{100\,000 \times r \times 5}{100}$$

$r = 8\%$

b $160\,000 - $140\,000 = $20\,000$

Question 66.3

a Using the Finance solver, with PMT to be solved, gives the monthly payment as $1288.95.

Compound Interest	
N	240
I%	7.5
PV	160000
PMT	−1288.94911
FV	0
P/Y	12
C/Y	12

b The constant a in the relation will be the multiplying factor associated with the monthly interest.

$$a = 1 + \frac{\frac{7.5}{12}}{100}$$

$$= 1.006\,25$$

The constant b is the amount of the repayment each month, so it will be a negative amount.
$b = -1288.95$

c Step 1: First enter the given information and calculate N, number of payments. N is 176 plus a payment of the balance.

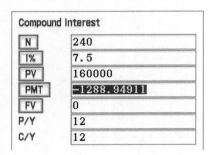

Compound Interest	
N	176.3267019
I%	7.5
PV	160000
PMT	−1500
FV	0
P/Y	12
C/Y	12

Step 2: Find what is owing after payment 176, then Dave must pay that plus the interest on that.

Compound Interest	
N	176
I%	7.5
PV	160000
PMT	−1500
FV	−488.0309435
P/Y	12
C/Y	12

$7.5\% \div 12 = 0.625\%$

Final payment = $488.03 + 0.625\% \times $488.03
(or 1.625\% × $488.03)
= $793.05 (to the nearest cent)

Dave would pay $1500 × 176 + $739.05
= $264\,739.05

This compares to $1288.95 × 240 = $309\,348

Saving = $309\,348 − $264\,739.05 = $44\,608.95

To the nearest $10, Dave makes a saving of $44\,610.

Question 66.4

a A depreciation of 15% is associated with a multiplying factor of $1 - 0.15 = 0.85$.

$100\,000 \times 0.85 = $85\,000$

b This is reducing balance depreciation. value after 5 years =

$100\,000 \times 0.85^5 = 44\,370.53$

($44\,371 to the nearest dollar)

c After five years, Dave has paid Fletch $40\,000 in interest.

Loss in value of the car =
$100\,000 - $44\,371 = $55\,629$

Total cost of Dave's car =
$40\,000 + $55\,629 = $95\,629$

67 Living it up

Question 67.1

a Use the perpetuity formula:

$Q = \dfrac{Pr}{100}$, where Q is the perpetuity amount,

P is the principal amount invested and r is the percentage interest per time period.

$r = \dfrac{4.5}{12} = 0.375\%$

$$Q = \dfrac{Pr}{100}$$
$$= \dfrac{780000 \times 0.375}{100}$$
$$= 2925; \$2925 \text{ per month}$$

b He will still have \$780 000 left after 15 years.

Question 67.2

a Using the Finance solver on CAS, he will have \$778 925 left after one payment.

Compound Interest	
N	1
I%	4.5
PV	−780000
PMT	4000
FV	778925
P/Y	12
C/Y	12

b The constant a in the relation will be the multiplying factor associated with the monthly interest.

$a = 1 + \dfrac{\frac{4.5}{12}}{100} = 1.00375$

The constant b is the amount of the payment each month so it will be a negative amount.

$b = -4000$

c After 120 months, he will still have \$617 462.07 of his original investment.

Compound Interest	
N	120
I%	4.5
PV	−780000
PMT	4000
FV	617462.0708
P/Y	12
C/Y	12

d Using the Finance solver, his investment will last for a total of 352 months.

Compound Interest	
N	351.0495635
I%	4.5
PV	−780000
PMT	4000
FV	0
P/Y	12
C/Y	12

Question 67.3

If inflation is at 2.5% per year, this represents a multiplying factor of $1 + \dfrac{2.5}{100} = 1.025$.

amount needed after 10 years

$= \$4000 \times 1.025^{10}$

$= \$5120.34$

Question 67.4

a

Compound Interest	
N	60
I%	4.5
PV	−780000
PMT	4000
FV	707818.5314
P/Y	12
C/Y	12

Using the Finance solver, after 5 years (60 months) receiving \$4000 per month, Sergio will have \$707 818.53 left.

b

Compound Interest	
N	202.2081571
I%	4.5
PV	−707818.5314
PMT	5000
FV	0
P/Y	12
C/Y	12

Using the Finance solver, the investment will last a further 203 months, where Sergio receives \$5000 per month.

Total number of months = 203 + 60
= 263 months

Question 67.5

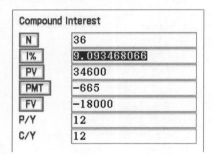

Compound Interest

N	180
I%	4.8
PV	−780000
PMT	6087.232594
FV	0
P/Y	12
C/Y	12

Using the Finance solver, if there is to be no money left after 15 years, then Sergio will receive $6087 per month.

68 Wheeler dealing

Question 68.1

$665 \times 36 = $23\,940$

Question 68.2

Use the Finance solver on CAS:

Compound Interest

N	36
I%	9.093468066
PV	34600
PMT	−665
FV	−18000
P/Y	12
C/Y	12

The interest rate would be 9.1%.

Question 68.3

a Payments over 5 years:
$60 \times $732 = $43\,920$

Balance after the deposit:
$34\,600 − $3460 = $31\,140$

Interest paid:
$43\,920 − $31\,140 = $12\,780$

b using the simple interest formula: $I = \dfrac{PrT}{100}$,
where $I = 12\,780$, $P = 31\,140$ and $T = 5$

$$12\,780 = \frac{31\,140 \times r \times 5}{100}$$
$$r = \frac{12\,780 \times 100}{31\,140 \times 5}$$
$$= 8.21\%$$

c Equivalent rate of interest $= \dfrac{2n}{n+1}$ flat rate;
where n is the number of payments

$$= \frac{2 \times 60}{61} \times 8.21 = 16.15\%$$

Question 68.4

a **i** Depreciation over 3 years
$= $34\,600 − $18\,000$
$= $16\,600$

Depreciation per year $= \dfrac{$16\,600}{3}$
$= 5533.33

As a percentage of the purchase price,
this is $\dfrac{5533.33}{34\,600} \times 100 = 15.99\%$.

  ii The constant a in the relation will have the value 1 as the interest is the same each year. The constant b is the amount of depreciation each year so it will be a negative amount.

$b = −5533.33$

$V_0 = 34\,600,\ V_{n+1} = V_n − 5533.33$

b **i** If r is the reducing balance rate of interest, then $PR^3 = 18\,000$, where $P = $34\,600$ and $R = 1 − \dfrac{r}{100}$.

$$34\,600 \times R^3 = 18\,000$$
$$R^3 = \frac{18\,000}{34\,600}$$
$$= 0.520\,23\ldots$$
$$R = \sqrt[3]{0.520\,23\ldots} = 0.8043$$
$$1 − \frac{r}{100} = 0.8043$$
$$\frac{r}{100} = 0.1957$$
$$r = 19.57\%$$

  ii The constant a in the relation will be the multiplying factor associated with the yearly interest.

$a = 1 − \dfrac{19.57}{100} = 0.8043$

The constant $b = 0$ as no amount is being added or subtracted each year.

$R_0 = 34\,600,\ R_{n+1} = 0.8043R_n$

c The car has travelled 75 000 kilometres and has depreciated $16 600.

$$\frac{$16\,600}{75\,000\text{ km}} = $0.2213 \text{ per kilometre}$$

The car depreciates at 22.13 cents per kilometre.

69 Original concepts

Question 69.1

a $6000 – $600 = $5400 depreciation

b Depreciation per year = $\dfrac{5400}{5}$ = $1080

Annual depreciation rate = $\dfrac{1080}{6000} \times 100$ = 18%

c The camera has been reduced to 10% of its original value, therefore 90% has been written off.

d Depreciation per year is $1080.

number of hours used = $\dfrac{1080}{4}$ = 270

e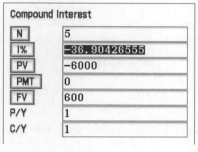

The depreciation rate would be 36.9% p.a.

Let the book value be $600 after 5 years.

$$600 = 6000 \left(1 - \frac{r}{100}\right)^5$$

$$0.1 = \left(1 - \frac{r}{100}\right)^5$$

$$1 - \frac{r}{100} = \sqrt[5]{0.1} = 0.630\,957\,344$$

$$\frac{r}{100} = 0.369; \; r = 36.9\%$$

f The reducing balance rate must be higher in order to give the same scrap value because, after the first year, it is continually calculated using a lower book value than that found when using flat rate depreciation.

g When using flat rate depreciation, the graph is linear, whereas for reducing balance it is non-linear (a decreasing curve).

Question 69.2

a $\dfrac{6.6}{4}$ = 1.65%

b

Quarter	Repayment ($)	Interest ($)	Principal reduction ($)	Balance ($)
1	3278.09	2640.00	638.09	159 361.91
2	3278.09	2629.46	648.63	158 713.28
3	3278.09	2618.77	659.32	158 053.96

c **i** Quarterly interest = $\dfrac{6.8}{4}$ = 1.7%

Repayments = 0.017 × 160 000 = $2720

ii An interest-only loan pays only the interest owing at the end of the time period.

Nothing is paid off the principal so he still owes $160 000.

70 Travelling

Question 70.1

a Use the compound interest formula:

$A = PR^n$ where $R = 1 + \dfrac{5.4}{12 \times 100} = 1.0045$;

$P = \$7500$ and $n = 36$

amount in the account $= 7500 \times 1.0045^{36}$
$= \$8815.75$

b   **i** The constant a in the relation will be the multiplying factor associated with the monthly interest.

$a = 1 + \dfrac{\frac{5.4}{12}}{100} = 1.0045$

The constant b is the amount of the additional payment each month, so it will be a positive amount.

$b = 100$

ii $R_1 = 1.0045 \times 7500 + 100 = \$7633.75$

$R_2 = 1.0045 \times 7633.75 + 100 = \$7768.10$

$R_3 = 1.0045 \times 7768.10 + 100 = \$7903.06$

iii Use the Finance solver on CAS:

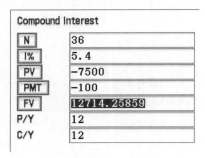

Compound Interest	
N	36
I%	5. 4
PV	−7500
PMT	−100
FV	12714.25859
P/Y	12
C/Y	12

Investments are considered 'outgoings', hence the negative signs.

She will have $12 714.26 in the account.

c Use the Finance solver on CAS:

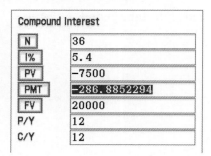

Compound Interest	
N	36
I%	5. 4
PV	−7500
PMT	−286.8852294
FV	20000
P/Y	12
C/Y	12

She will need to invest $286.89 per month to have $20 000 accumulated after 3 years.

d Use the Finance solver on CAS:

Compound Interest	
N	36
I%	5. 4
PV	−7500
PMT	−286.7553977
FV	20000
P/Y	12
C/Y	365

If interest is credited daily, then she will need to make payments of $286.76 per month.

Question 70.2

a Use the Finance solver on CAS:

Compound Interest	
N	12
I%	5. 4
PV	−7500
PMT	−300
FV	11605.62577
P/Y	12
C/Y	12

She will have $11 605.63 in the account after 12 months.

b Use the Finance solver on CAS:

Compound Interest	
N	21.92733873
I%	6. 4
PV	−11605.63
PMT	−300
FV	20000
P/Y	12
C/Y	12

total number of months $= 21.9 + 12$
$= 33.9$ months

After 34 months she will have $20 000 in the account.

c Use the Finance solver on CAS:

Compound Interest	
N	24
I%	6. 4
PV	−11605.63
PMT	−300
FV	20845.29357
P/Y	12
C/Y	12

After 12 months with an interest rate of 5.4% p.a. and 24 months at an interest rate of 6.4% p.a., she will have $20 845.29 in the account.

d Interest earned
$$= \$20\,845.29 - (\$7500 + 36 \times \$300)$$
$$= \$2545.29$$

Question 70.3

Solving $20\,000 = (7500 + P) \times 1.004\,5^{36}$ for P,

$$\frac{20000}{1.0045^{36}} = 7500 + P$$

$$17\,015.01 = 7500 + P$$

$$9515.01 = P$$

If Rebecca deposits \$9515.01 in her account now (in addition to the \$7500), then she will have \$20\,000 in 3 years' time.

71 School funding

Question 71.1

a Monthly rate of interest $= \dfrac{4.8}{12} = 0.4\%$

b 0.4% per month is associated with a multiplying factor of $1 + \dfrac{0.4}{100} = 1.004$

$$1000 \times 1.004 = 1004$$
$$1004 \times 1.004 = 1008.016$$
$$1008.016 \times 1.004 = 1012.048$$

The amount after 3 months is \$1012.05.

c Multiplying factor $= 1.004 = b$

No further payments are made, so c is 0.

u_0 is the initial amount in the account, so $a = 1000$.

$$u_0 = 1000, \ u_{n+1} = 1.004u_n$$

Question 71.2

a $1050 \times 1.004 + 100 = 1154.2$; \$1154.20 in the account after the first month.

b The constant a in the relation will be the multiplying factor associated with the monthly interest.

$$a = 1 + \frac{\frac{4.8}{12}}{100} = 1.004$$

The constant b is the amount of the additional payment each month, so it will be a positive amount.

$$b = 100$$

$$A_0 = 1050, \ A_{n+1} = 1.004A_n + 100$$

c $A_1 = 1.004 \times A_0 + 100$
  $= 1.004 \times 1050 + 100$
  $= 1154.20$

$A_2 = 1.004 \times A_1 + 100$
  $= 1.004 \times 1154.20 + 100$
  $= 1258.82$

$A_3 = 1.004 \times A_2 + 100$
  $= 1.004 \times 1258.82 + 100$
  $= 1363.86$

$A_4 = 1.004 \times A_3 + 100$
  $= 1.004 \times 1363.86 + 100$
  $= 1469.31$

d

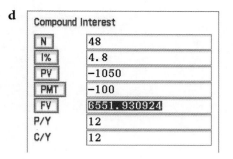

Compound Interest	
N	48
I%	4.8
PV	-1050
PMT	-100
FV	6551.930924
P/Y	12
C/Y	12

There is \$6551.93 in the account after 48 months.

Question 71.3

Use the Finance solver on CAS:

Compound Interest	
N	84
I%	4.8
PV	-6552
PMT	-510.4165081
FV	60000
P/Y	12
C/Y	12

Adam's parents will need to deposit \$510.41 in the account so that they have \$60\,000 in the account at the end of 7 years.

Question 71.4

Substitute in the formula for compound increase:
$A = PR^n$ where $P = 10\,000$, $R = 1.025$ and $n = 7$

$$A = 10\,000 \times 1.025^7 = 11\,886.86$$

The inflated value after 7 years is \$11\,887.

Question 71.5

Use the Finance solver on CAS:

There are 6 payments of $11 234.64, one at the beginning of each year, but there are 12 compounding periods per year.

The payment time needs to be changed to the beginning of the year, payments per year to 1 and compounding periods per year to 12.

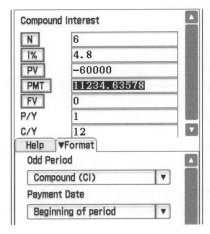

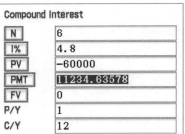

72 Martine's car

Question 72.1

a Using CAS, Martine will need to repay her parents $4593.99 per year.

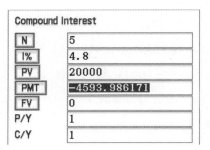

b

Payment number	Payment ($)	Interest ($)	Principal reduction ($)	Balance ($)
1	4000.00	960.00	3040.00	16 960.00
2	4000.00	814.08	3185.92	13 774.08
3	4000.00	661.16	3338.84	10 435.24
4	4000.00	500.89	3499.11	6 936.13
5	7269.06	332.93	6936.13	0.00

c Martine will need to repay her parents a total of $7269.06 at the end of the fifth year.

d Over the 5 years, Martine has paid her parents a total of $3269.06 in interest. This can be found by

adding the interest column, or alternatively

using total payments – loan
= 23 269.06 – 20 000
= $3269.06

Question 72.2

a A depreciation of 24% is associated with a multiplying factor of $1 - \dfrac{24}{100} = 0.76$.

$\$19\,850 \times 0.76 = \$15\,086$

b A depreciation of 15% is associated with a multiplying factor of $1 - \dfrac{15}{100} = 0.85$.

The value after the next four years will be $\$15\,086 \times 0.85^4 = \$7874.99 \approx \$7875$

Question 72.3

a $\$24\,620 - \$7000 = \$17\,620$

b Interest paid = $36 \times 600 - 17\,620 = 3980$
$\$3980$ in interest on a loan of $\$17\,620$.

c Flat-rate loan. The interest paid is $3980 over 3 years, which is $1326.67 per year.

If the interest is a flat rate, then the interest rate is $\dfrac{1326.67}{17\,620} \times 100 = 7.53\%$.

This is the quoted interest rate, so it must be a flat rate.

73 VCAA 2016 Exam 2

Question 73.1

a $\$15\,000$

b $V_1 = 1.04 \times 15\,000 = \$15\,600$
$V_2 = 1.04 \times 15\,600 = \$16\,224$

c $(1.04 - 1) \times 100 = 4\%$

d **i** $V_n = 1.04^n \times 15\,000$

   ii $\$22\,203.66$

Question 73.2

a $\dfrac{38\,000 - 16\,000}{8} = \$2750$

b $C_0 = 38\,000,\ C_{n+1} = C_n - 2750$

c $8 \times 5000 = 40\,000\,\text{km}$
$\dfrac{22\,000}{40\,000} = \$0.55$

Question 73.3

a **i**

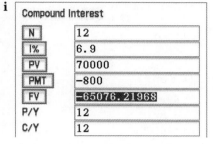

   $\$65\,076.22$

   ii $12 \times 800 - (70\,000 - 65\,076.22) = \$4676.22$

b FV after 3 years is $\$54\,151.60$.

Calculate new PV after lump sum to pay off in 3 years:

PV = $\$25\,947.58$

Difference = $L = \$28\,204$

74 VCAA 2017 Exam 2

Question 74.1

a $V_0 = 75\,000$
$V_1 = 75\,000 - 3375 = 71\,625$
$V_2 = 71\,625 - 3375 = 68\,250$

b **i** $\$3375$

   ii $\dfrac{3375}{75\,000} \times 100 = 4.5\%$

c $(1 - 0.943) \times 100 = 5.7\%$

Question 74.2

a $1.015 \times 200 = \$203$

b $A_0 = 428,\ A_{n+1} = 1.015 A_n$

c $A_4 = 454.26$

Interest = $454.26 - 428 = \$26.26$

Question 74.3

a   5.2% of 36 000 = 18 720

$$\frac{18\,720}{12} = \$1560$$

b

Compound Interest

N	48
I%	3.8
PV	−360000
PMT	−500
FV	444872.9445
P/Y	12
C/Y	12

FV after 4 years = $444 872.94

Compound Interest

N	24
I%	3.8
PV	−444872.9445
PMT	−805.6505059
FV	500000
P/Y	12
C/Y	12

Payment required = $805.65

75 VCAA 2017 NHT Exam 2

Question 75.1

a $180

b $V_1 = 3000 - 180 = 2820$

$V_2 = 2820 - 180 = 2640$

c   6 years

d   **i** $\dfrac{2539.20}{2760} = 0.92 = 8\%$ depreciation

ii $S_0 = 3000, S_{n+1} = 0.92 S_n$

Question 75.2

a $P_1 = 1.056 P_0$

$1584 = 1.056 A$

$A = \dfrac{1584}{1.056} = 1500$

b $1865.29

c $Q_n = 2080.05 \times 1.0046^n$

$$1 + \frac{\frac{5.52}{12}}{100}$$

Question 75.3

a

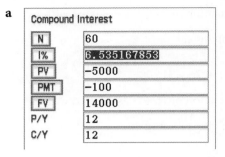

Compound Interest

N	60
I%	6.535167853
PV	−5000
PMT	−100
FV	14000
P/Y	12
C/Y	12

6.54%

b   After 36 months: $9964.63

After further 24 months: $11 276.52

76 VCAA 2018 NHT Exam 2

Question 76.1

a $5000

b $V_2 - V_0 = \$512.50$

c $V_0 = 5000, V_{n+1} = 1.05 V_n$

d $R^2 \times 5000 = 6000$

Interest rate = 9.5%

Question 76.2

a   **i** $S_4 = \$5032$

ii $\dfrac{867}{8500} \times 100 = 10.2\%$

b $8500 \times 0.92^4 = \$6089.34$

c $4500 = 8500 \times \left(1 - \dfrac{d}{100}\right)^4$

$d = 14.7\%$

Question 76.3

a $\dfrac{12.9}{100} \times 10\,000 \times \dfrac{1}{12} = \$107.50$

b

Compound Interest

N	36
I%	12.9
PV	10000
PMT	−250.0000619
FV	−3776.15
P/Y	12
C/Y	12

$250

c $420.40

A Finance solver approach could have been

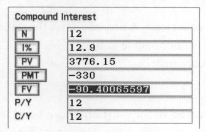

Hence the last payment of $330 must be increased by $90.40 to fully repay the loan.

UNIT 4

Chapter 3

Matrices

Multiple-choice solutions

1 E

This matrix has three rows and one column; order 3×1.

A column matrix has order $m \times 1$, i.e. there is only one column.

2 C

The order of the matrix is: number of rows × number of columns

3 D

Matrix multiplication AB is possible if the number of columns in A is equal to the number of rows in B. There are two columns in A and two rows in B, so multiplication is possible.

4 B

An $m \times n$ matrix multiplied by an $n \times p$ matrix gives an $m \times p$ matrix.

A 1×2 matrix multiplied by a 2×3 matrix gives a 1×3 matrix.

5 E

Matrix addition is possible if the two matrices to be added are of the same order.

6 D

x_{ij} is the element in the ith row and the jth column

x_{31} is the element in the 3rd row and the 1st column and has value -3. This is the only correct alternative.

7 C

The only definition that applies to this matrix is a triangular matrix; zeros below the leading diagonal. Element $x_{23} = 3$; the element in the second row and third column.

8 A

The inverse of $\begin{bmatrix} 1 & 0 \\ 0 & 1 \end{bmatrix}$ is $\begin{bmatrix} 1 & 0 \\ 0 & 1 \end{bmatrix}$.

This matrix is called the identity or unit matrix. It is also a square and a diagonal matrix.

9 C

$$X - 2Y = \begin{bmatrix} 1 & 0 \\ a & b \end{bmatrix} - 2\begin{bmatrix} 2 & 2 \\ 0 & 0 \end{bmatrix}$$

$$= \begin{bmatrix} 1-4 & 0-4 \\ a-0 & b-0 \end{bmatrix}$$

$$= \begin{bmatrix} -3 & -4 \\ a & b \end{bmatrix}$$

10 C

BA is the product of a 1×3 matrix and a 3×1 matrix and so the resulting product will be a 1×1 matrix.

The products BC, AC and CB cannot be found because the orders of the matrices are not correct for multiplication. CA can be found; it will be a 2×1 matrix.

11 B

$2A + 3B$

$$= \begin{bmatrix} 2 \times 3 & 2 \times 1 \\ 2 \times 4 & 2 \times 6 \end{bmatrix} + \begin{bmatrix} 3 \times 2 & 3 \times (-1) \\ 3 \times (-2) & 3 \times 0 \end{bmatrix}$$

$$= \begin{bmatrix} 6 & 2 \\ 8 & 12 \end{bmatrix} + \begin{bmatrix} 6 & -3 \\ -6 & 0 \end{bmatrix}$$

$$= \begin{bmatrix} 12 & -1 \\ 2 & 12 \end{bmatrix}$$

12 A

$$\begin{bmatrix} a & b \\ c & d \end{bmatrix} + \begin{bmatrix} -a & -b \\ -c & -d \end{bmatrix} = \begin{bmatrix} 0 & 0 \\ 0 & 0 \end{bmatrix}$$

13 A

The matrix A^2 exists only if matrix A is a square matrix.

Option **A** is a 1×1 matrix; the only square matrix.

14 D

$$C^2 = \begin{bmatrix} 2 & 4 \\ 1 & -1 \end{bmatrix} \times \begin{bmatrix} 2 & 4 \\ 1 & -1 \end{bmatrix}$$

$$= \begin{bmatrix} 2 \times 2 + 4 \times 1 & 2 \times 4 + 4 \times (-1) \\ 1 \times 2 + (-1) \times 1 & 1 \times 4 + (-1) \times (-1) \end{bmatrix}$$

$$= \begin{bmatrix} 8 & 4 \\ 1 & 5 \end{bmatrix}$$

15 C

If a matrix is of the form $\begin{bmatrix} a & b \\ c & d \end{bmatrix}$, then the determinant is $ad - bc$. The determinant of each matrix is:

A: $(1 \times 4) - (2 \times 3) = 4 - 6 = -2$

B: $(5 \times 0) - (7 \times 3) = 0 - 21 = -21$

C: $(-1 \times 4) - (-3 \times 3) = -4 + 9 = 5$

D: $(0 \times 0) - (2 \times 3) = 0 - 6 = -6$

E: $(7 \times 4) - (9 \times 3) = 28 - 27 = 1$

16 D

BA is a 2×4 matrix, which means there are two rows of four elements; $2 \times 4 = 8$ elements in the product matrix.

17 E

The inverse of $\begin{bmatrix} a & b \\ c & d \end{bmatrix}$ is $\begin{bmatrix} \dfrac{d}{\Delta} & \dfrac{-b}{\Delta} \\ \dfrac{-c}{\Delta} & \dfrac{a}{\Delta} \end{bmatrix}$,

where $\Delta = ad - bc$.

So the inverse of $\begin{bmatrix} w & x \\ y & z \end{bmatrix}$ is

$$\begin{bmatrix} \dfrac{z}{wz - xy} & \dfrac{-x}{wz - xy} \\ \dfrac{-y}{wz - xy} & \dfrac{w}{wz - xy} \end{bmatrix}.$$

18 E

A determinant can only be found for a square matrix.

Matrix A has 2 rows and 3 columns. If A is multiplied by a 3×2 matrix, the product will be a 2×2 matrix. Another 2×3 matrix can be subtracted from A because it has the same order. It is not possible to find A^2 because A is not a square matrix.

19 C

Only matrices of the same order can be added or subtracted, so the expression containing A and D (both 2×2) is the only calculation that is possible.

20 D

Matrices A and B are both 2×2 matrices, so it is possible to multiply, add, or square the matrices.

A^{-1} is the inverse of matrix A and this only exists if the determinant is not zero.

The determinant of A is $(3 \times 4) - (6 \times 2) = 12 - 12 = 0$, so the inverse cannot be found.

21 D

$$A + I = \begin{bmatrix} 1 & 2 \\ 3 & 4 \end{bmatrix} + \begin{bmatrix} 1 & 0 \\ 0 & 1 \end{bmatrix}$$

$$= \begin{bmatrix} 2 & 2 \\ 3 & 5 \end{bmatrix}$$

$$\neq \begin{bmatrix} 1 & 2 \\ 3 & 4 \end{bmatrix}$$

22 E

A singular matrix has a determinant of zero. Checking the determinant of each option:

A: $(3 \times 0) - (-1 \times -3) = 0 - 3 = -3$

B: $(1 \times 1) - (1 \times 0) = 1 - 0 = 1$

C: $(2 \times -2) - (-2 \times -2) = -4 - 4 = -8$

D: $(1 \times 25) - (6 \times 4) = 25 - 24 = 1$

E: $(3 \times -3) - (-3 \times 3) = -9 + 9 = 0$

23 D

The determinant of the matrix is $2 \times (-2) - 1 \times (-1) = -3$.

We swap the elements on the leading diagonal and change the sign of the other elements to find the inverse.

The inverse is: $-\dfrac{1}{3} \begin{bmatrix} -2 & 1 \\ -1 & 2 \end{bmatrix} = \begin{bmatrix} \dfrac{2}{3} & -\dfrac{1}{3} \\ \dfrac{1}{3} & -\dfrac{2}{3} \end{bmatrix}$

24 C

$A = 2B$ gives: $\begin{bmatrix} a & b \\ c & d \end{bmatrix} = \begin{bmatrix} 4 & 6 \\ 0 & 2 \end{bmatrix}$

So: $a = 4$, $b = 6$, $c = 0$ and $d = 2 = \frac{1}{2}a$;

$c = 2$ is not true.

25 C

The element in the second row, third column, is found by multiplying the second row of the first matrix by the third column of the second matrix:

$2 \times (-1) + 4 \times 2 + (-1) \times 3 = 3$

26 E

Matrix $x = \begin{bmatrix} x_{11} & x_{12} \\ x_{21} & x_{22} \\ x_{31} & x_{33} \end{bmatrix}$

$\begin{bmatrix} 1+1 = 2 & 1+2 = 3 \\ 2+1 = 3 & 2+2 = 4 \\ 3+1 = 4 & 3+2 = 5 \end{bmatrix} = \begin{bmatrix} 2 & 3 \\ 3 & 4 \\ 4 & 5 \end{bmatrix}$

27 A

$\frac{1}{4}\begin{bmatrix} 4 & 3 \\ 2 & 8 \end{bmatrix} = \frac{1}{2}\begin{bmatrix} a & b \\ 1 & 4 \end{bmatrix}$

Hence: $\begin{bmatrix} 1 & \frac{3}{4} \\ \frac{1}{2} & 2 \end{bmatrix} = \begin{bmatrix} \frac{a}{2} & \frac{b}{2} \\ \frac{1}{2} & 2 \end{bmatrix}$

giving: $\frac{a}{2} = 1 \Rightarrow a = 2$

$\frac{b}{2} = \frac{3}{4} \Rightarrow b = \frac{6}{4} = \frac{3}{2}$

28 B

The rows have been changed in the matrix XM and this is produced by a permutation matrix.

The permutation $\begin{bmatrix} 0 & 0 & 1 \\ 1 & 0 & 0 \\ 0 & 1 & 0 \end{bmatrix}$ of the

identity matrix $\begin{bmatrix} 1 & 0 & 0 \\ 0 & 1 & 0 \\ 0 & 0 & 1 \end{bmatrix}$ will change

the first row to the second row, the second row to the third row and the third row to the first row, as seen in the matrix XM.

29 D

$B + B^2 = \begin{bmatrix} -1 & 0 \\ 1 & 1 \end{bmatrix} + \begin{bmatrix} -1 & 0 \\ 1 & 1 \end{bmatrix} \times \begin{bmatrix} -1 & 0 \\ 1 & 1 \end{bmatrix}$

$= \begin{bmatrix} -1 & 0 \\ 1 & 1 \end{bmatrix} + \begin{bmatrix} 1 & 0 \\ 0 & 1 \end{bmatrix}$

$= \begin{bmatrix} 0 & 0 \\ 1 & 2 \end{bmatrix}$

30 D

The determinant of $\begin{bmatrix} 2 & -3 \\ -1 & x \end{bmatrix}$ is:

$2 \times x - (-1 \times -3) = 2x - 3$

The inverse is $\frac{1}{2x - 3}\begin{bmatrix} x & 3 \\ 1 & 2 \end{bmatrix}$;

swapping the elements on the leading diagonals and changing the sign on the others.

31 D

$$\begin{matrix} I & \times & D & = & D \\ (\ldots \times \ldots) & \times & (1 \times 3) & = & (1 \times 3) \end{matrix}$$

Same

The order of I will have to be 1×1 because I is a square matrix.

32 D

To multiply two matrices, the number of columns in the first matrix must be the same as the number of rows in the second matrix.

33 E

The inverse of a matrix does not exist if the determinant is zero.

The determinant of $\begin{bmatrix} a & 8 \\ 5 & 4 \end{bmatrix}$ is

$a \times 4 - 5 \times 8 = 4a - 40$.

If $4a - 40 = 0$, then $a = 10$.

34 A

Use CAS:

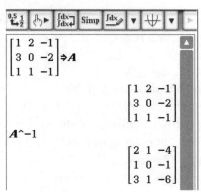

35 E

Replacing the '1's with arrows, the only routes in both directions are C to E.

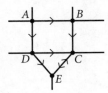

36 B

With four teams, there are
$\dfrac{n(n-1)}{2} = \dfrac{4 \times 3}{2} = 6$ matches. The only matrix that shows 6 matches is option **B**. Also the 1s are in the correct places for the results given.

37 B

Team C defeated teams B and E.

Team B defeated teams D and E. Team E defeated team A.

So, team C has a two-step dominance over teams A, D and E.

Alternatively, find the square of the matrix that will give the second stage wins.

38 D

If M is the given dominance matrix, then M^2 is the two-step dominance matrix.

using CAS:

$$M + M^2 = \begin{bmatrix} 0 & 2 & 1 & 1 & 2 \\ 2 & 0 & 1 & 1 & 2 \\ 1 & 1 & 0 & 1 & 2 \\ 2 & 2 & 2 & 0 & 2 \\ 1 & 1 & 1 & 0 & 0 \end{bmatrix}$$

$$\text{dominance vector} = \begin{bmatrix} 6 \\ 6 \\ 5 \\ 8 \\ 3 \end{bmatrix}$$

Ranking is D first, A and B equal second, C fourth and E last.

39 E

If lunch is with Craig on Monday (today) then lunch is with Edgar on Tuesday (tomorrow). Similarly, Wednesday is with Daniel, Thursday is with Betty and Friday is with Angela.

40 B

There are six games played so there should be only six '1's in the matrix (because a '1' represents a win). The only matrix with six '1's is option **B**. The direction of the edge joining each pair of vertices can to be observed, showing option **B** to be correct.

41 A

Summing the dominance value for each team does not rank the teams so, if A is the adjacency matrix, find $A + A^2$ to find the sum of one-step and two-step dominances.

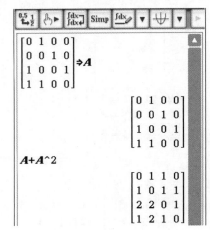

Summing the rows of the matrix $A + A^2$ gives the total dominance scores as follows: 2 to team A, 3 to team B, 5 to team C and 4 to team D.

So the rankings, in order, are C, D, B, A.

42 B

Option **B** is not true:

$$PQP = P \times (QP)$$
and: $\qquad P^2Q = P \times (PQ)$

In general: $QP \neq PQ$

43 E

$AA^{-1} = I$ so:

$(AA^{-1})^{-1} = I^{-1} = I$

Not matrix A^{-1}, as in option **E**.

44 B

Try squaring each of the options:

$A : \begin{bmatrix} 2 & 2 \\ 0 & 2 \end{bmatrix}\begin{bmatrix} 2 & 2 \\ 0 & 2 \end{bmatrix} = \begin{bmatrix} 2 & 8 \\ 0 & 4 \end{bmatrix}$ Not correc

$B : \begin{bmatrix} 2 & 1 \\ 0 & 2 \end{bmatrix}\begin{bmatrix} 2 & 1 \\ 0 & 2 \end{bmatrix} = \begin{bmatrix} 4 & 4 \\ 0 & 4 \end{bmatrix}$ Correct

45 E

The given information in equation form is:

$7x + 8y = 816.5$

$10x - 3y = 293.5$

As a matrix equation, this is:

$$\begin{bmatrix} 7 & 8 \\ 10 & -3 \end{bmatrix} \begin{bmatrix} x \\ y \end{bmatrix} = \begin{bmatrix} 816.5 \\ 293.5 \end{bmatrix}$$

46 C

$$\begin{bmatrix} -3 \\ 4 \end{bmatrix} \begin{bmatrix} a & b \end{bmatrix} = \begin{bmatrix} -3a & -3b \\ 4a & 4b \end{bmatrix}$$

$$= \begin{bmatrix} 6 & x \\ y & 12 \end{bmatrix}$$

Equating matrices gives:

$-3a = 6 \Rightarrow a = -2$

$4b = 12 \Rightarrow b = 3$

$-3b = x$, so $x = -3 \times 3 = -9$

$4a = y$, so $y = 4 \times (-2) = -8$

47 C

$$\begin{bmatrix} 5 & 2 \\ 8 & 4 \end{bmatrix} B = 2 \begin{bmatrix} 1 & 0 \\ 0 & 1 \end{bmatrix}$$

So $\begin{bmatrix} 5 & 2 \\ 8 & 4 \end{bmatrix} \times \dfrac{1}{2}B = \begin{bmatrix} 1 & 0 \\ 0 & 1 \end{bmatrix}$

$\dfrac{1}{2}B$ will be the inverse of $\begin{bmatrix} 5 & 2 \\ 8 & 4 \end{bmatrix}$.

$$\frac{1}{2}B = \frac{1}{4} \begin{bmatrix} 4 & -2 \\ -8 & 5 \end{bmatrix}$$

$$B = \frac{1}{2} \begin{bmatrix} 4 & -2 \\ -8 & 5 \end{bmatrix}$$

48 A

$$AB = 9 \begin{bmatrix} 1 & 0 & 0 \\ 0 & 1 & 0 \\ 0 & 0 & 1 \end{bmatrix}$$

So: $A \times \dfrac{1}{9}B = \begin{bmatrix} 1 & 0 & 0 \\ 0 & 1 & 0 \\ 0 & 0 & 1 \end{bmatrix}$

Hence, the inverse of A is $\dfrac{1}{9}B$.

49 C

In a transition matrix, the columns (or rows) must sum to one. The figure 34% refers to the initial state and is not part of the transition matrix.

$$\begin{array}{cc} & \text{Now} \\ & \text{Tri-Hi \quad Others} \\ \text{Next} \begin{array}{c} \text{Tri-Hi} \\ \text{Others} \end{array} & \begin{bmatrix} 0.7 & 0.1 \\ 0.3 & 0.9 \end{bmatrix} \end{array}$$

50 C

The steady-state proportion is found by finding a high power of the transition matrix:

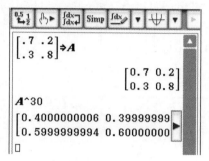

51 C

After three transition periods, the state will be $T^3 S_0$, where T is the transition matrix and S_0 is the initial state matrix.

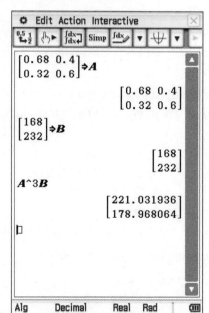

52 E

The transition matrix will be:

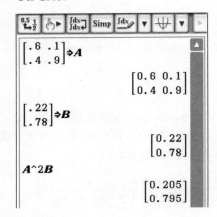

Now

		Apex	Other
Next	Apex	0.6	0.1
	Other	0.4	0.9

10% of those who rent from another company will rent from Apex the next month.

The initial state matrix will be the column matrix:

Apex $\begin{bmatrix} 0.22 \\ 0.78 \end{bmatrix}$
Other

53 B

Apex's percentage share of the car rental market after 2 months will be

$$\begin{bmatrix} 0.6 & 0.1 \\ 0.4 & 0.9 \end{bmatrix}^2 \begin{bmatrix} 0.22 \\ 0.78 \end{bmatrix}.$$

On CAS:

$$\begin{bmatrix} .6 & .1 \\ .4 & .9 \end{bmatrix} \Rightarrow A$$

$$\begin{bmatrix} 0.6 & 0.1 \\ 0.4 & 0.9 \end{bmatrix}$$

$$\begin{bmatrix} .22 \\ .78 \end{bmatrix} \Rightarrow B$$

$$\begin{bmatrix} 0.22 \\ 0.78 \end{bmatrix}$$

$$A{\wedge}2B$$

$$\begin{bmatrix} 0.205 \\ 0.795 \end{bmatrix}$$

54 A

The steady-state proportion is found by finding a high power of the transition matrix.

$$A{\wedge}50$$

$$\begin{bmatrix} 0.2 & 0.2 \\ 0.8 & 0.8 \end{bmatrix}$$

55 A

Site C is the only migration site to Site A.

20% from Site C migrate to Site A.

20% of 2800 = 560

56 D

Looking at the diagram, the mutton birds migrate to sites B and D and then never leave. Eventually they will all be there.

57 D

If x is the number of birds at each site in 2007, in 2008:

$0.35x$ (from A) + $0.15x$ (from C) + x (at B)
= 6000

$1.5x = 6000$
$x = 4000$

Total number in 2007 = $4x$
= 16 000

58 B

Today

		Fine	Showers
Next day	Fine	0.35	0.4
	Showers	0.65	0.6

65% of the days that are fine today will have showers tomorrow.

59 C

The chance of it being fine in 2 days' time can be found using T^2S_0, where T is the transition matrix:

$$\begin{bmatrix} .35 & .4 \\ .65 & .6 \end{bmatrix} \Rightarrow A$$

$$\begin{bmatrix} 0.35 & 0.4 \\ 0.65 & 0.6 \end{bmatrix}$$

$$\begin{bmatrix} 1 \\ 0 \end{bmatrix} \Rightarrow B$$

$$\begin{bmatrix} 1 \\ 0 \end{bmatrix}$$

$$A{\wedge}2B$$

$$\begin{bmatrix} 0.3825 \\ 0.6175 \end{bmatrix}$$

60 B

Finding a high power of the transition matrix gives the steady-state proportions:

$$A{\wedge}50$$

$$\begin{bmatrix} 0.380952381 & 0.380952381 \\ 0.619047619 & 0.619047619 \end{bmatrix}$$

61 C

If $S_3 = TS_2$ then $T^{-1}S_3 = T^{-1}TS_2$

$= S_2$

$$= \begin{bmatrix} 1100 \\ 900 \end{bmatrix}$$

62 D

The Leslie matrix is a square matrix with the first row as the birth rate and the 2nd and 3rd giving the survival rate for each time period.

63 C

The first row of the Leslie matrix refers to reproduction rates and the second and third rows refer to survival rates. So, 0.47 refers to the survival rate of age group 0–<3 over the first 3 years and, hence, 47% of this age group survive the first 3 years and go on to the next group.

64 A

Multiply the initial state matrix $\begin{bmatrix} 43 \\ 56 \\ 32 \end{bmatrix}$ by the Leslie matrix will give the numbers in each age group after 3 years (one time period)

$$\begin{bmatrix} 0.36 & 0.8 & 0.25 \\ 0.47 & 0 & 0 \\ 0 & 0.52 & 0 \end{bmatrix}\begin{bmatrix} 43 \\ 56 \\ 32 \end{bmatrix} = \begin{bmatrix} 68.3 \\ 20.2 \\ 29.1 \end{bmatrix}, \text{ so}$$

68 individuals (rounded) in the 0–<3 after 3 years.

65 E

12 years is 4 time periods so we are comparing the initial population $\begin{bmatrix} 43 \\ 56 \\ 32 \end{bmatrix}$ with the matrix

$$\begin{bmatrix} 0.36 & 0.8 & 0.25 \\ 0.47 & 0 & 0 \\ 0 & 0.52 & 0 \end{bmatrix}^4 \times \begin{bmatrix} 43 \\ 56 \\ 32 \end{bmatrix} = \begin{bmatrix} 39 \\ 21 \\ 12 \end{bmatrix}$$

Both populations have decreased in the 0–<6 and 6–9 age groups.

Extended-answer solutions

66 Car hire

Question 66.1

a **i** This is a product of a 3 × 3 matrix and a 3 × 1 matrix, so the product will be the 3 × 1 matrix:

$$\begin{bmatrix} 6 \\ 13 \\ 10 \end{bmatrix}$$

ii This product matrix represents the sum of the rows of matrix A.

b **i** $\begin{bmatrix} 3 & 2 & 1 \\ 4 & 3 & 6 \\ 5 & 2 & 3 \end{bmatrix} \times \begin{bmatrix} 1 & 0 & 0 \\ 0 & 2 & 0 \\ 0 & 0 & 3 \end{bmatrix} = \begin{bmatrix} 3 & 4 & 3 \\ 4 & 6 & 18 \\ 5 & 4 & 9 \end{bmatrix}$

ii In the product matrix, the elements in column 1 of A are multiplied by 1, the elements in column 2 of A are multiplied by 2, and the elements in column 3 of A are multiplied by 3.

Question 66.2

a $\begin{bmatrix} 50 & 45 & 40 & 37 & 35 \\ 60 & 55 & 50 & 47 & 45 \\ 70 & 65 & 60 & 57 & 55 \\ 80 & 75 & 70 & 67 & 65 \end{bmatrix}$

b $\begin{bmatrix} 1 & 0 & 0 & 0 & 0 \\ 0 & 2 & 0 & 0 & 0 \\ 0 & 0 & 3 & 0 & 0 \\ 0 & 0 & 0 & 4 & 0 \\ 0 & 0 & 0 & 0 & 5 \end{bmatrix}$

c $\begin{bmatrix} 50 & 45 & 40 & 37 & 35 \\ 60 & 55 & 50 & 47 & 45 \\ 70 & 65 & 60 & 57 & 55 \\ 80 & 75 & 70 & 67 & 65 \end{bmatrix} \times \begin{bmatrix} 1 & 0 & 0 & 0 & 0 \\ 0 & 2 & 0 & 0 & 0 \\ 0 & 0 & 3 & 0 & 0 \\ 0 & 0 & 0 & 4 & 0 \\ 0 & 0 & 0 & 0 & 5 \end{bmatrix} = \begin{bmatrix} 50 & 90 & 120 & 148 & 175 \\ 60 & 110 & 150 & 188 & 225 \\ 70 & 130 & 180 & 228 & 275 \\ 80 & 150 & 210 & 268 & 325 \end{bmatrix}$

Question 66.3

a $\begin{bmatrix} 58 & 47 & 62 \\ 114 & 121 & 141 \\ 127 & 108 & 127 \\ 59 & 61 & 73 \\ 43 & 23 & 38 \end{bmatrix}$

b $\begin{bmatrix} 1 \\ 1 \\ 1 \end{bmatrix}$; $\begin{bmatrix} 58 & 47 & 62 \\ 114 & 121 & 141 \\ 127 & 108 & 127 \\ 59 & 61 & 73 \\ 43 & 23 & 38 \end{bmatrix} \times \begin{bmatrix} 1 \\ 1 \\ 1 \end{bmatrix} = \begin{bmatrix} 167 \\ 376 \\ 362 \\ 193 \\ 104 \end{bmatrix}$

c $\begin{bmatrix} 50 & 90 & 120 & 148 & 175 \end{bmatrix}$

d $\begin{bmatrix} 50 & 90 & 120 & 148 & 175 \end{bmatrix} \times \begin{bmatrix} 167 \\ 376 \\ 362 \\ 193 \\ 104 \end{bmatrix} = \begin{bmatrix} 132\,394 \end{bmatrix}$

Total revenue from the hiring of small cars for the three months is $132 394.

67 Changing cars

Question 67.1

a $\begin{bmatrix} 1 & 1 & 1 \end{bmatrix} \times \begin{bmatrix} 3 & 2 & 1 \\ 4 & 3 & 6 \\ 5 & 2 & 3 \end{bmatrix} = \begin{bmatrix} 12 & 7 & 10 \end{bmatrix}$

b The product matrix represents the sum of each of the columns of matrix A.

Question 67.2

a $\begin{bmatrix} 58 & 60 & 35 & 11 \\ 114 & 108 & 132 & 27 \\ 127 & 148 & 96 & 52 \\ 59 & 96 & 112 & 48 \\ 43 & 29 & 89 & 53 \end{bmatrix}$

b $\begin{bmatrix} 1 & 1 & 1 & 1 & 1 \end{bmatrix}$

c $\begin{bmatrix} 401 & 441 & 464 & 191 \end{bmatrix}$

Question 67.3

a

		Last rental	
		A-A	B-B
This rental	A-A	0.9	0.4
	B-B	0.1	0.6

b $T = \begin{bmatrix} 0.9 & 0.4 \\ 0.1 & 0.6 \end{bmatrix}$

c $S_0 = \begin{bmatrix} 0.55 \\ 0.45 \end{bmatrix}$

d **i** $TS_0 = \begin{bmatrix} 0.9 & 0.4 \\ 0.1 & 0.6 \end{bmatrix} \times \begin{bmatrix} 0.55 \\ 0.45 \end{bmatrix} = \begin{bmatrix} 0.675 \\ 0.325 \end{bmatrix}$

A-A has 67.5% of the market after one transition period.

ii $T^5 S_0 = \begin{bmatrix} 0.9 & 0.4 \\ 0.1 & 0.6 \end{bmatrix}^5 \times \begin{bmatrix} 0.55 \\ 0.45 \end{bmatrix} = \begin{bmatrix} 0.792 \\ 0.208 \end{bmatrix}$

A-A has 79.2% of the market after five transition periods.

e $T^6 S_0 = T^7 S_0 = \begin{bmatrix} 0.80 \\ 0.20 \end{bmatrix}$

After 6 transition periods.

f The steady-state percentage for company A-A is 80%.

68 Fuel for thought

Question 68.1

a $Q = \begin{bmatrix} 16\,800 & 10\,890 & 5660 \\ 15\,400 & 12\,630 & 8320 \\ 14\,250 & 9980 & 7160 \end{bmatrix}$

b $X = \begin{bmatrix} u \\ g \\ d \end{bmatrix}$

c $T = \begin{bmatrix} 32\,839.55 \\ 35\,137.85 \\ 31\,103.45 \end{bmatrix}$

d $QX = T$

$$\begin{bmatrix} 16\,800 & 10\,890 & 5660 \\ 15\,400 & 12\,630 & 8320 \\ 14\,250 & 9\,980 & 7160 \end{bmatrix} \begin{bmatrix} u \\ g \\ d \end{bmatrix}$$

$$= \begin{bmatrix} 32\,839.55 \\ 35\,137.85 \\ 31\,103.45 \end{bmatrix}$$

Question 68.2

a

	Last week (%)		
This week (%)		Light rail	Private car	Other
	Light rail	80	10	50
	Private car	10	70	20
	Other	10	20	30

b $\begin{bmatrix} 0.8 & 0.1 & 0.5 \\ 0.1 & 0.7 & 0.2 \\ 0.1 & 0.2 & 0.3 \end{bmatrix}$

c $\begin{bmatrix} 0.4 \\ 0.5 \\ 0.1 \end{bmatrix}$

d **i** $S_1 = TS_0$

$$= \begin{bmatrix} 0.8 & 0.1 & 0.5 \\ 0.1 & 0.7 & 0.2 \\ 0.1 & 0.2 & 0.3 \end{bmatrix} \times \begin{bmatrix} 0.4 \\ 0.5 \\ 0.1 \end{bmatrix}$$

$$= \begin{bmatrix} 0.42 \\ 0.41 \\ 0.17 \end{bmatrix}$$

42.0% of commuters are using the light-rail after one week.

ii $S_5 = T^5 S_0 = \begin{bmatrix} 0.8 & 0.1 & 0.5 \\ 0.1 & 0.7 & 0.2 \\ 0.1 & 0.2 & 0.3 \end{bmatrix}^5 \times \begin{bmatrix} 0.4 \\ 0.5 \\ 0.1 \end{bmatrix}$

$$= \begin{bmatrix} 0.5268 \\ 0.3078 \\ 0.1654 \end{bmatrix}$$

52.7% of commuters are using the light-rail after five weeks.

> **Note**
> Matrix multiplication is done on CAS.

e Steady state is reached when $TS_{n+1} \approx TS_n$ or is found by finding a high power of T:

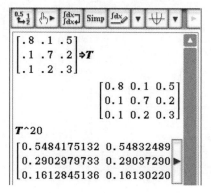

The steady-state percentage of commuters using light-rail is 55%, to the nearest percentage.

69 Matrix manoeuvres

Question 69

a $2X$ is the matrix in which each element in X is multiplied by 2.

So: $2X = \begin{bmatrix} 6 & 0 & 4 \\ 2 & -2 & 8 \\ -4 & 2 & 0 \end{bmatrix}$

b X^2 is the product:

$$\begin{bmatrix} 3 & 0 & 2 \\ 1 & -1 & 4 \\ -2 & 1 & 0 \end{bmatrix} \times \begin{bmatrix} 3 & 0 & 2 \\ 1 & -1 & 4 \\ -2 & 1 & 0 \end{bmatrix}$$

$$= \begin{bmatrix} 5 & 2 & 6 \\ -6 & 5 & -2 \\ -5 & -1 & 0 \end{bmatrix}$$

c $X^T = \begin{bmatrix} 3 & 1 & -2 \\ 0 & -1 & 1 \\ 2 & 4 & 0 \end{bmatrix}$

Rows become columns in the transpose of a matrix.

d
$$\begin{bmatrix} \frac{1}{2} & 0 & 0 \\ 0 & \frac{1}{2} & 0 \\ 0 & 0 & \frac{1}{2} \end{bmatrix} = \frac{1}{2}\begin{bmatrix} 1 & 0 & 0 \\ 0 & 1 & 0 \\ 0 & 0 & 1 \end{bmatrix}$$

$$= \frac{1}{2} \times \text{identity matrix}$$

So:
$$\begin{bmatrix} 3 & 0 & 2 \\ 1 & -1 & 4 \\ -2 & 1 & 0 \end{bmatrix} \times \begin{bmatrix} \frac{1}{2} & 0 & 0 \\ 0 & \frac{1}{2} & 0 \\ 0 & 0 & \frac{1}{2} \end{bmatrix}$$

$$= \frac{1}{2}\begin{bmatrix} 3 & 0 & 2 \\ 1 & -1 & 4 \\ -2 & 1 & 0 \end{bmatrix} \times \begin{bmatrix} 1 & 0 & 0 \\ 0 & 1 & 0 \\ 0 & 0 & 1 \end{bmatrix}$$

$$= \frac{1}{2}\begin{bmatrix} 3 & 0 & 2 \\ 1 & -1 & 4 \\ -2 & 1 & 0 \end{bmatrix}$$

$$= \begin{bmatrix} 1.5 & 0 & 1 \\ 0.5 & -0.5 & 2 \\ -1 & 0.5 & 0 \end{bmatrix}$$

e The matrix A must be a 3×3 matrix, otherwise the equation cannot work.

$$2X - A = \begin{bmatrix} 5 & 4 & -1 \\ 2 & 1 & 3 \\ -3 & 4 & -5 \end{bmatrix} \text{ means:}$$

$$A = 2X - \begin{bmatrix} 5 & 4 & -1 \\ 2 & 1 & 3 \\ -3 & 4 & -5 \end{bmatrix}$$

$$= \begin{bmatrix} 6 & 0 & 4 \\ 2 & -2 & 8 \\ -4 & 2 & 0 \end{bmatrix} - \begin{bmatrix} 5 & 4 & -1 \\ 2 & 1 & 3 \\ -3 & 4 & -5 \end{bmatrix}$$

$$= \begin{bmatrix} 1 & -4 & 5 \\ 0 & -3 & 5 \\ -1 & -2 & 5 \end{bmatrix}$$

f **i** Using CAS to find the inverse of X, enter the elements of the matrix X into CAS:

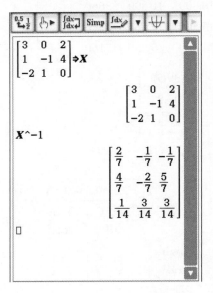

The inverse is

$$\begin{bmatrix} \frac{2}{7} & -\frac{1}{7} & -\frac{1}{7} \\ \frac{4}{7} & -\frac{2}{7} & \frac{5}{7} \\ \frac{1}{14} & \frac{3}{14} & \frac{3}{14} \end{bmatrix}$$

ii $\dfrac{1}{14}\begin{bmatrix} 4 & -2 & -2 \\ 8 & -4 & 10 \\ 1 & 3 & 3 \end{bmatrix}$

g $DX = \begin{bmatrix} 14 & 0 & 0 \\ 0 & 14 & 0 \\ 0 & 0 & 14 \end{bmatrix} = 14 \times I$

So $\left(\dfrac{1}{14}D\right)X = I$

Hence $\dfrac{1}{14}D = X^{-1} = \dfrac{1}{14}\begin{bmatrix} 4 & -2 & -2 \\ 8 & -4 & 10 \\ 1 & 3 & 3 \end{bmatrix}$

So $D = \begin{bmatrix} 4 & -2 & -2 \\ 8 & -4 & 10 \\ 1 & 3 & 3 \end{bmatrix}$

70 Mixes

Question 70.1

a $M = \dfrac{1}{100}\begin{bmatrix} 50 & 25 & 25 \\ 30 & 45 & 25 \\ 40 & 25 & 35 \end{bmatrix}$ or $\begin{bmatrix} 0.5 & 0.25 & 0.25 \\ 0.3 & 0.45 & 0.25 \\ 0.4 & 0.25 & 0.35 \end{bmatrix}$

b $C = \begin{bmatrix} 6 \\ 10 \\ 5 \end{bmatrix}$

c $MC = P$

Question 70.2

a Copper balance: $20x + 0.25y + 0.5z + 0.2w = 2400 \times 1$

Lead balance: $x + 55y + z + 0.5w = 2400 \times 8$

Zinc balance: $x + y + 45z + w = 2400 \times 15$

Iron balance: $5x + 2y + 2z + 20w = 2400 \times 10$

b $\begin{bmatrix} 20 & 0.25 & 0.5 & 0.2 \\ 1 & 55 & 1 & 0.5 \\ 1 & 1 & 45 & 1 \\ 5 & 2 & 2 & 20 \end{bmatrix} \times \begin{bmatrix} x \\ y \\ z \\ w \end{bmatrix} = 2400 \times \begin{bmatrix} 1 \\ 8 \\ 15 \\ 10 \end{bmatrix}$

71 Pizza planning

Question 71.1

a **i** $\begin{bmatrix} 5 & 3 & -2 \\ 4 & 2 & 1 \\ 7 & -1 & 4 \end{bmatrix} \times \begin{bmatrix} 1 \\ 1 \\ 1 \end{bmatrix} = \begin{bmatrix} 6 \\ 7 \\ 10 \end{bmatrix}$

ii The elements in the product matrix are the sum of the elements in the rows of matrix A.

b **i** $\begin{bmatrix} 1 & 1 & 1 \end{bmatrix} \times \begin{bmatrix} 5 & 3 & -2 \\ 4 & 2 & 1 \\ 7 & -1 & 4 \end{bmatrix}$

$= \begin{bmatrix} 16 & 4 & 3 \end{bmatrix}$

ii The elements in the product matrix are the sum of the elements in the columns of matrix A.

Question 71.2

a $P = \begin{bmatrix} 28 & 35 & 28 & 36 & 37 & 34 & 45 \\ 36 & 37 & 47 & 32 & 36 & 38 & 47 \\ 41 & 42 & 40 & 35 & 51 & 56 & 61 \\ 30 & 33 & 39 & 33 & 78 & 76 & 83 \end{bmatrix}$

b $A = \begin{bmatrix} 1 \\ 1 \\ 1 \\ 1 \\ 1 \\ 1 \\ 1 \end{bmatrix}$

c PA

$= \begin{bmatrix} 28 & 35 & 28 & 36 & 37 & 34 & 45 \\ 36 & 37 & 47 & 32 & 36 & 38 & 47 \\ 41 & 42 & 40 & 35 & 51 & 56 & 61 \\ 30 & 33 & 39 & 33 & 78 & 76 & 83 \end{bmatrix} \times \begin{bmatrix} 1 \\ 1 \\ 1 \\ 1 \\ 1 \\ 1 \\ 1 \end{bmatrix}$

$= \begin{bmatrix} 243 \\ 273 \\ 326 \\ 372 \end{bmatrix}$

There were 243 small, 273 medium, 326 large and 372 family-size pizzas sold in a week.

d $B = \begin{bmatrix} 1 & 1 & 1 & 1 \end{bmatrix}$

e $BP = \begin{bmatrix} 1 & 1 & 1 & 1 \end{bmatrix} \times$

$$\begin{bmatrix} 28 & 35 & 28 & 36 & 37 & 34 & 45 \\ 36 & 37 & 47 & 32 & 36 & 38 & 47 \\ 41 & 42 & 40 & 35 & 51 & 56 & 61 \\ 30 & 33 & 39 & 33 & 78 & 76 & 83 \end{bmatrix} =$$

$$\begin{bmatrix} 135 & 147 & 154 & 136 & 202 & 204 & 236 \end{bmatrix}$$

The numbers of pizzas sold were 135 on Monday, 147 on Tuesday, 154 on Wednesday, 136 on Thursday, 202 on Friday, 204 on Saturday, and 236 on Sunday.

Question 71.3

a $\begin{bmatrix} 32 & 110 & 8 & 200 \\ 50 & 175 & 10 & 240 \\ 72 & 250 & 12 & 280 \\ 98 & 345 & 15 & 320 \end{bmatrix} \times \begin{bmatrix} 1 \\ 1 \\ 1 \\ 1 \end{bmatrix} = \begin{bmatrix} 350 \\ 475 \\ 614 \\ 778 \end{bmatrix}$

A small pizza costs 350 cents to make (production cost); a medium pizza costs 475 cents; a large pizza costs 614 cents, and a family-size pizza costs 778 cents.

b $\begin{bmatrix} 9.5 \\ 11.5 \\ 13.5 \\ 17.5 \end{bmatrix} - \dfrac{1}{100} \begin{bmatrix} 350 \\ 475 \\ 614 \\ 778 \end{bmatrix} = \begin{bmatrix} 6 \\ 6.75 \\ 7.36 \\ 9.72 \end{bmatrix}$

$6 profit is made on a small pizza; $6.75 on a medium pizza; $7.36 on a large pizza and $9.72 on a family-size pizza.

c $\begin{bmatrix} 243 \\ 273 \\ 326 \\ 372 \end{bmatrix}$ is a 4 × 1 matrix, so transposing this

to a 1 × 4 matrix and multiplying by the profit matrix gives:

$$\begin{bmatrix} 243 & 273 & 326 & 372 \end{bmatrix} \times \begin{bmatrix} 6 \\ 6.75 \\ 7.36 \\ 9.72 \end{bmatrix}$$

$$= \begin{bmatrix} 9315.95 \end{bmatrix}$$

The profit for the week is $9315.95.

72 Playing to win

Question 72.1

Number of matches played $= \dfrac{n(n-1)}{2}$

$$= \dfrac{6 \times 5}{2}$$

$$= 15$$

Question 72.2

a   Team F defeated teams B and D, so these are the two teams that team F played.

b   Teams A and F won two of the matches that they actually played. Teams B and D won three of the games that they actually played.

Question 72.3

a   The dominance vector is $R = \begin{bmatrix} 3 \\ 3 \\ 2 \\ 3 \\ 2 \\ 2 \end{bmatrix}$.

b   The three teams that are ranked equal first are A, B and D.

c $M^2 = \begin{bmatrix} 0 & 2 & 0 & 1 & 1 & 2 \\ 1 & 0 & 2 & 0 & 3 & 2 \\ 0 & 2 & 0 & 1 & 0 & 1 \\ 0 & 1 & 1 & 0 & 2 & 3 \\ 1 & 1 & 1 & 2 & 0 & 0 \\ 2 & 0 & 2 & 1 & 1 & 0 \end{bmatrix}$

d M^2 tells us the two-step dominance. For example, if A defeated C and C defeated E, then A has a two-step dominance over E.

e

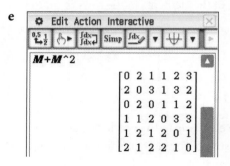

Dominance vector is $\begin{bmatrix} 9 \\ 11 \\ 6 \\ 10 \\ 7 \\ 8 \end{bmatrix}$.

f   Teams are ranked B first, D second, A third, F fourth, E fifth and C last.

9780170465335

Question 72.4

a
$$\begin{bmatrix} 0 & 0 & 1 & 0 & 1 \\ 1 & 0 & 1 & 1 & 0 \\ 0 & 0 & 0 & 0 & 1 \\ 1 & 0 & 1 & 0 & 1 \\ 0 & 1 & 0 & 0 & 0 \end{bmatrix}$$

b
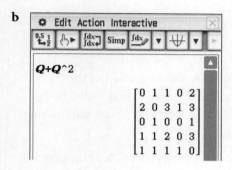

$Q+Q^2$

$$\begin{bmatrix} 0 & 1 & 1 & 0 & 2 \\ 2 & 0 & 3 & 1 & 3 \\ 0 & 1 & 0 & 0 & 1 \\ 1 & 1 & 2 & 0 & 3 \\ 1 & 1 & 1 & 1 & 0 \end{bmatrix}$$

c dominance vector = $\begin{bmatrix} 4 \\ 9 \\ 2 \\ 7 \\ 4 \end{bmatrix}$

d Teams are ranked: *B* first, *D* second, *A* and *E* equal third, *C* last.

73 Read all about it

Question 73.1

a i
$$\begin{bmatrix} 123 & 108 & 48 & 12 \\ 136 & b & 43 & 15 \\ 115 & 133 & 36 & 11 \\ 145 & 128 & 45 & 18 \end{bmatrix}$$

ii
$$\begin{bmatrix} 348.00 \\ 377.60 \\ 345.30 \\ 405.50 \end{bmatrix}$$

b
$$\begin{bmatrix} 1.20 \\ 1.00 \\ 1.30 \\ 2.50 \end{bmatrix}$$

c If
$$\begin{bmatrix} 123 & 108 & 48 & 12 \\ 136 & b & 43 & 15 \\ 115 & 133 & 36 & 11 \\ 145 & 128 & 45 & 18 \end{bmatrix}\begin{bmatrix} 1.20 \\ 1.00 \\ 1.30 \\ 2.50 \end{bmatrix}=\begin{bmatrix} 348.00 \\ 377.60 \\ 345.30 \\ 405.50 \end{bmatrix}$$

then

$$136 \times 1.2 + b \times 1 + 43 \times 1.3 + 15 \times 2.5 = 377.60$$
$$b = 121$$

Question 73.2

$$\begin{bmatrix} 130 & 111 & 45 & 11 \\ 143 & 126 & 40 & 14 \\ 121 & 135 & 42 & 12 \\ 136 & 128 & 39 & 14 \end{bmatrix}\begin{bmatrix} a \\ b \\ c \\ d \end{bmatrix}=\begin{bmatrix} 370.85 \\ 403.20 \\ 384.60 \\ 395.55 \end{bmatrix}$$

Question 73.3

a $S_0 = \begin{bmatrix} 136 \\ 128 \\ 39 \\ 14 \end{bmatrix}$

b
$$\begin{bmatrix} 0.96 & 0.04 & 0.11 & 0.06 \\ 0.01 & 0.95 & 0.03 & 0 \\ 0.02 & 0.01 & 0.85 & 0.01 \\ 0.01 & 0 & 0.01 & 0.93 \end{bmatrix}\times\begin{bmatrix} 136 \\ 128 \\ 39 \\ 14 \end{bmatrix}$$

$$=\begin{bmatrix} 140.81 \\ 124.13 \\ 37.29 \\ 14.77 \end{bmatrix}$$

$$=\begin{bmatrix} 141 \\ 124 \\ 37 \\ 15 \end{bmatrix}$$

(to the nearest whole numbers)

There would be 141 of type *A* sold, 124 of type *B* sold, 37 of type *C* sold, and 15 of type *D* sold.

c Finding a high power of the transition matrix, *T*, gives the long term, or steady-state, proportions.

T^{100} gives the steady-state proportions and numbers of each type that can be expected to be sold:

Type *A*: $0.614 \times 320 = 196$

Type *B*: $0.183 \times 320 = 59$

Type *C*: $0.101 \times 320 = 32$

Type *D*: $0.102 \times 320 = 33$

74 Training

Question 74.1

a $68x + 143y + 85z = 921.90$

b Using Monday, Tuesday and Wednesday:

$$\begin{bmatrix} 68 & 143 & 85 \\ 53 & 146 & 72 \\ 72 & 139 & 64 \end{bmatrix} \begin{bmatrix} x \\ y \\ z \end{bmatrix} = \begin{bmatrix} 921.90 \\ 821.00 \\ 872.30 \end{bmatrix}$$

Any three of the days could be included in this equation. There is an oversupply of information.

Question 74.2

a

	Today	
	On time	Late
Tomorrow On time	90%	95%
Tomorrow Late	10%	5%

b $T = \begin{bmatrix} 0.9 & 0.95 \\ 0.1 & 0.05 \end{bmatrix}$

c $S_0 = \begin{bmatrix} 1 \\ 0 \end{bmatrix} \begin{matrix} \text{On time} \\ \text{Late} \end{matrix}$

d i If S_0 is the state on Monday, then $S_1 = TS_0$ is the state on Tuesday.

$$TS_0 = \begin{bmatrix} 0.9 & 0.95 \\ 0.1 & 0.05 \end{bmatrix} \begin{bmatrix} 1 \\ 0 \end{bmatrix}$$

$$= \begin{bmatrix} 0.9 \times 1 + 0.95 \times 0 \\ 0.1 \times 1 + 0.05 \times 0 \end{bmatrix}$$

$$= \begin{bmatrix} 0.9 \\ 0.1 \end{bmatrix}$$

There is a 90% chance that the train will be on time on Tuesday.

ii S_4 is the state on Friday.

$$S_4 = T^4 S_0 = \begin{bmatrix} 0.9048 \\ 0.0952 \end{bmatrix} \text{ using CAS.}$$

There is a 90.5% chance that the train will be on time on Friday.

e $S_5 = T^5 S_0 = \begin{bmatrix} 0.9048 \\ 0.0952 \end{bmatrix} \approx S_4$

Hence we have reached a steady state.

There is a 90.5% chance of a train being on time.

75 Island hopping

Question 75.1

a

This year

$$T = \begin{matrix} & A & B & C & L & \\ & \begin{bmatrix} 0.7 & 0 & 0 & 0 \\ 0.1 & 0.8 & 0 & 0 \\ 0.15 & 0.05 & 0.92 & 0 \\ 0.05 & 0.15 & 0.08 & 1 \end{bmatrix} & \begin{matrix} A \\ B \\ C \\ L \end{matrix} \end{matrix} \text{ Next year}$$

The figures are the proportion of the population that moves to each of the other locations in the following year. Columns must sum to 1.

b 0.92 refers to the fact that 92% of the people on island C remain on island C the next year.

c

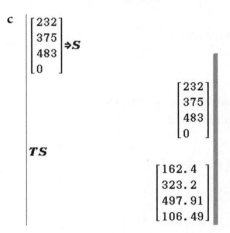

i $S_{2017} = \begin{bmatrix} 162.4 \\ 323.2 \\ 497.91 \\ 106.49 \end{bmatrix}$

ii Pre-multiplying the previous question by matrix T:

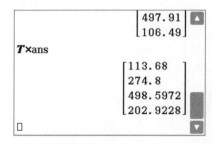

In 2018, the expected number of people on island A is 114 (to the nearest whole number).

iii Executing the same process another two times will give the values for 2020.

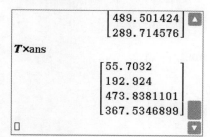

The number of people on island B in 2020 is 193 (to the nearest whole number).

iv Executing the same process another six times will give the values for 2026:

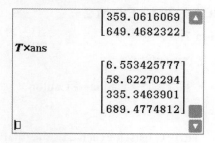

The total number of people to leave the islands at the beginning of 2026 is 689.

v The number of people on island C in the long term is 0.

T^{100}×ans

$$\begin{bmatrix} 2.119690173\text{E-}15 \\ 1.327660978\text{E-}8 \\ 0.08742761455 \\ 1089.912572 \end{bmatrix}$$

Question 75.2

a $Q = \begin{bmatrix} 50 \\ 40 \\ 20 \\ 0 \end{bmatrix} \begin{matrix} A \\ B \\ C \\ L \end{matrix}$

b i Using CAS to find $TS + Q$:

$TS+Q$

$$\begin{bmatrix} 212.4 \\ 363.2 \\ 517.91 \\ 106.49 \end{bmatrix}$$

The number of inhabitants of island A in 2017 is 212 (to the nearest whole number).

ii Using CAS: Pre-multiply the previous question by T, then add Q. Executing this process three times gives the numbers for 2020.

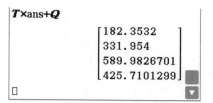

The expected number of inhabitants of island B at the beginning of 2020 is 332.

iii The total number of inhabitants at the beginning of 2020 is $182 + 332 + 590 = 1104$.

c The population of island A is decreasing but by a lesser number each year

2016: 232; 2017: 212; 2018:199; 2019: 189 2020:182; 2021: 178;

(the population will stabilise at 167 in the long term).

The population of island C is increasing but by a lesser number each year

2017: 518; 2018: 546; 2019:570; 2020: 590; 2021: 607;

(the population will stabilise at 740 in the long term).

76 VCAA 2011 Exam 2

Question 76.1

a i Birds eat lizards.

ii Nothing eats birds.

b

$$Z = \begin{matrix} & I & B & L & F \\ \begin{bmatrix} 0 & 1 & 1 & 1 \\ 0 & 0 & 0 & 0 \\ 0 & 1 & 0 & 0 \\ 0 & 1 & 1 & 0 \end{bmatrix} & \begin{matrix} I \\ B \\ L \\ F \end{matrix} \end{matrix}$$

Question 76.2

a $K = \begin{bmatrix} 99\,500 & 20 & 25 & 240 \end{bmatrix}$

b $0.05 \times 400 = 20$

c $M = \begin{bmatrix} 285 \end{bmatrix} = \begin{bmatrix} 99\,500 & 20 & 25 & 240 \end{bmatrix} \times \begin{bmatrix} 0 \\ 1 \\ 1 \\ 1 \end{bmatrix}$

d The total number of birds, lizards and frogs killed.

Question 76.3

a $32 + 64 = 96$

b $W_1 = \begin{bmatrix} 0 & 2 \\ 0.25 & 0.5 \end{bmatrix} \times \begin{bmatrix} 32 \\ 64 \end{bmatrix} = \begin{bmatrix} 128 \\ 40 \end{bmatrix}$

c **i** Use CAS to find W_2 then W_3:

$$W_2 = \begin{bmatrix} 80 \\ 52 \end{bmatrix}$$

$$W_3 = \begin{bmatrix} 104 \\ 46 \end{bmatrix}$$

104 juvenile ducks

46 adult ducks

The numbers of juvenile and adult ducks are plotted on the graph below.

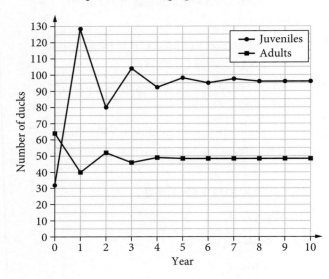

ii $B^{10} W_0 = \begin{bmatrix} 96 \\ 48 \end{bmatrix}$

Numbers do not change after this so $96 + 48 = 144$.

d $\begin{bmatrix} 0 & 1 \\ 0.25 & 0.5 \end{bmatrix}^4 \begin{bmatrix} 32 \\ 64 \end{bmatrix} = \begin{bmatrix} 28 \\ 23 \end{bmatrix}$ total of 51

$\begin{bmatrix} 0 & 1 \\ 0.25 & 0.5 \end{bmatrix}^5 \begin{bmatrix} 32 \\ 64 \end{bmatrix} = \begin{bmatrix} 23 \\ 19 \end{bmatrix}$ total of 42

So during the 5th year.

e Solve

$$\begin{bmatrix} 0 & 1 \\ 0.25 & 0.5 \end{bmatrix}^2 \begin{bmatrix} a \\ b \end{bmatrix} = \begin{bmatrix} 100 \\ 50 \end{bmatrix}$$

Pre-multiply by

$$\left(\begin{bmatrix} 0 & 1 \\ 0.25 & 0.5 \end{bmatrix}^2 \right)^{-1}$$

$$\begin{bmatrix} a \\ b \end{bmatrix} = \begin{bmatrix} 8 & -8 \\ -2 & 4 \end{bmatrix} \begin{bmatrix} 100 \\ 4 \end{bmatrix} = \begin{bmatrix} 400 \\ 0 \end{bmatrix}$$

So 400 juveniles and 0 adults.

77 VCAA 2012 Exam 2

Question 77.1

a Anvil, Dantel

b Anvil $\to$ Berga $\to$ Dantel $\to$ Cantor

c $G = \begin{bmatrix} 1 & 2 & 1 & 1 \end{bmatrix}$

d The matrix G lists, for each city, the total number of direct flight connections from that city to another city in the network. It refers specifically to the number of direct flights out of each of the four cities to another city in the network, not the inward flights or just 'connections'.

Question 77.2

a **i** $C = \begin{bmatrix} 1 & 3 & 2 \\ 3 & 9 & 6 \end{bmatrix}$

ii 1 3 3 9 2 6

b If $C = BA$ then $B^{-1}C = B^{-1}BA = A$.

Question 77.3

a 70% of A + 80% of 0 + 90% of P

$= 70 + 160 + 45$

$= 275\%$

b 100% of staff who leave any year remain off staff the next year i.e. no staff ever return once they leave.

c **i** Using CAS, $TS_{2011} = \begin{bmatrix} 70 \\ 170 \\ 65 \\ 45 \end{bmatrix}$

ii $T^2 S_{2011} = \begin{bmatrix} 49 \\ 143 \\ 76 \\ 83 \end{bmatrix}$

So 143 in 2013.

iii $T^{10} S_{2011} \Rightarrow 2021$; the year when the number of operators is 29. Use CAS.

iv $T^{50} = \begin{bmatrix} 0 \\ 0 \\ 0 \\ 1 \end{bmatrix}$

d 182; Using CAS

$S_{2012} = \begin{bmatrix} 100 \\ 190 \\ 75 \\ 45 \end{bmatrix}$; $S_{2013} = T S_{2012} + A = \begin{bmatrix} 100 \\ 182 \\ 96.5 \\ 91.5 \end{bmatrix}$

78 VCAA 2014 Exam 2

Question 78.1

a 4×2

b 1850 males

c The total number of adult females who live in this city.

d $V \times P$ is a $(4 \times 2) \times (2 \times 1)$ product. The number of columns in V equals the number of rows in P. VP has order 4×1.

e $w = 1360 \times 0.45 + 1460 \times 0.55 = 1415$

f The sum of the elements in VP

$= 1415 + 1812 + 988 + 1806$
$= 6021$

Question 78.2

a **i** 20% of the voters move from C to A

   ii 5% to B + 20% to C. So 25%

b To Mr Broad (B): 5% per month from A + 40% per month from C

$= 5/100 \times 6000 + 40/100 \times 2160$
$= 1164$

c **i** Using CAS, $S_3 = \begin{bmatrix} 4900 \\ 4634 \\ 2466 \end{bmatrix}$

   ii The number of voters preferring each candidate in March.

d Use CAS to find $S_6 = T^5 S_1$

$= \begin{bmatrix} 4334 \\ 5303 \\ 2363 \end{bmatrix}$

B wins with 5303 votes.

Question 78.3

a The 10% of B voters that would have gone to C are now added 5% to B and 5% to A. Required percentage is 50%

b Voters at the end of June

$= T_1 \times T^4 S_1$

$= \begin{bmatrix} 5549 \\ 6451 \\ 0 \end{bmatrix} \begin{matrix} A \\ B \\ C \end{matrix}$

6451 votes went to Mr Broad.

79 VCAA 2015 Exam 2

Question 79.1

a $20 + 60 + 40 = 120$

b **i** $Q = \begin{bmatrix} 5 & ⑩ & 3 & 2 \\ 15 & 30 & ⑨ & 6 \\ 10 & 20 & 6 & 4 \end{bmatrix} \begin{matrix} B \\ I \\ A \end{matrix}$

with column headers $D \quad E \quad F \quad G$

   ii 30 students

c **i** CQ gives the amount paid.

   ii $CQ = \begin{bmatrix} 850 & 1700 & 510 & 340 \end{bmatrix}$

with column headers $D \quad E \quad F \quad G$

Geoff is paid \$340 each week.

Question 79.2

a All advanced level students remain at this level after assessment.

b **i** $\begin{bmatrix} 10 \\ 58 \\ 52 \end{bmatrix}$ Using CAS to find $T_1 S_0$

   ii Advanced level students have increased by 12 students from 40 to 52.

Question 79.3

a

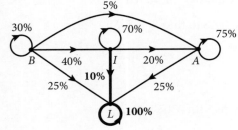

b Using CAS to calculate

$$T_2R_0 = \begin{bmatrix} 6 \\ 50 \\ 43 \\ 21 \end{bmatrix}$$

21 of the 120 students leave so 17.5% leave.

c $T_2{}^2R_0 = \begin{bmatrix} 1.8 \\ 37.4 \\ 42.55 \\ 38.25 \end{bmatrix} \begin{matrix} B \\ I \\ A \\ L \end{matrix}$

After two assessment 43 advanced level students remain.

d $T_2{}^5R_0 = \begin{bmatrix} 0 \\ 13 \\ 30 \\ 76 \end{bmatrix} \begin{matrix} B \\ I \\ A \\ L \end{matrix}$

After 5 assessment there are less than 50 students.

e Using CAS

$$R_3 = \begin{bmatrix} 6.1 \\ 34.48 \\ 48.13 \\ 58.29 \end{bmatrix}$$

34.48 students are present. From matrix T_2, 20% are expected to become advanced level. 20% of 34.48 = 6.9 Hence 7 students.

Question 80

a S_0 is the 4 × 1 matrix of the number of individuals in each age group.

From the table $S_0 = \begin{bmatrix} 32 \\ 25 \\ 18 \\ 14 \end{bmatrix}$.

b

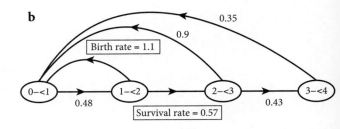

c **i** $S_1 = LS_0$

$$= \begin{bmatrix} 0 & 1.1 & 0.9 & 0.35 \\ 0.48 & 0 & 0 & 0 \\ 0 & 0.57 & 0 & 0 \\ 0 & 0 & 0.43 & 0 \end{bmatrix} \begin{bmatrix} 32 \\ 25 \\ 18 \\ 14 \end{bmatrix} = \begin{bmatrix} 49 \\ 15 \\ 14 \\ 8 \end{bmatrix}$$

(figures to the nearest whole number)

ii 15 individuals in the group 1–<2

iii $S_5 = L^5S_0 = \begin{bmatrix} 29 \\ 15 \\ 10 \\ 4 \end{bmatrix}$

Initial population total
= 32 + 25 + 18 + 14 = 89

Population after 5 years
= 29 + 15 + 10 + 4 = 58

Hence, a difference of
89 − 58 = 31 fewer individuals.

Chapter 4

Networks and decision mathematics

Multiple-choice solutions

1 D

The degree of each vertex is the number of edges connecting to it. The most common degree is 3.

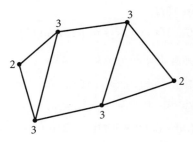

2 B

The graph is not directed as there are no arrows on the edges. The graph is not weighted as the edges do not have weights. It is not complete as every vertex does not join every other vertex. The graph is not simple as there are two edges joining vertices Q and R.

The graph is planar as it can be drawn so that no edges intersect.

3 C

A spanning tree is a connected graph with no circuits. D and E are not connected; A and B have circuits.

4 C

To use Euler's formula, the graph must be connected and planar. Option **C** is not a connected graph.

5 B

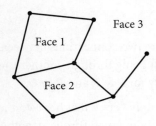

6 A

A subgraph contains some or all of the vertices from the original graph and some or all of the edges. An edge on a subgraph must be between two existing vertices and there can be no new edges (which eliminates options **B**, **C**, **D** and **E**).

7 E

Drawing the edges to represent a boundary in common:

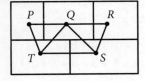

8 B

A complete graph has an edge from each vertex to every other vertex.

3 vertices, 3 edges   4 vertices, 6 edges   5 vertices, 10 edges

9 B

A connected graph has no isolated vertices.

n vertices, $n - 1$ edges.

3 vertices, 2 edges   4 vertices, 3 edges   5 vertices, 4 edges

10 A

If x is the number of faces, then substituting in Euler's formula gives:

$$v + f = e + 2$$
$$2x + x = 16 + 2$$
$$3x = 18$$
$$x = 6$$

11 D

For an Euler circuit to exist, the degree of all the vertices in a graph must be even. Only graph **D** has this property.

12 D

In a simple graph, the sum of the degrees of the graph is equal to twice the number of edges of the graph. Therefore, a simple graph with 7 edges will have 14 as the sum of the degrees of the vertices.

13 C

Loops are represented by the numbers, other than zero, along the leading diagonal of a matrix.

There are two loops on vertex '4'.

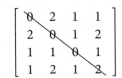

$$\begin{bmatrix} 0 & 2 & 1 & 1 \\ 2 & 0 & 1 & 2 \\ 1 & 1 & 0 & 1 \\ 1 & 2 & 1 & 2 \end{bmatrix}$$

14 B

The graph is connected: there are no isolated vertices. The graph is planar: there are no edges that intersect. The graph is not a tree (there are circuits) and not complete (where each vertex would be connected to every other vertex).

15 A

A connected graph has no isolated vertices.

10 vertices, 9 edges

16 E

Using Euler's formula:

$v + f = e + 2$

where $v = 10$ (vertices) and $f = 10$ (faces)

$10 + 10 = e + 2$

$e = 18$

17 A

An Euler trail can be taken if all but two of the vertices are even. The path will start and finish at the 'odd' vertices. So an Euler trail could start at A and finish at B, or vice versa.

18 D

A Hamiltonian cycle starts and finishes at the same vertex (which eliminates options **A** and **B**) and goes through each vertex once only (which eliminates options **C** and **E**).

19 D

A is not a spanning tree because it has a circuit. Adding the weightings on the edges of the other options gives **D** as the minimum.

20 C

For an Euler circuit to exist, all vertices must have an even degree. Vertices T and U are of odd degree, so an edge joining these two vertices would mean that all vertices are of even degree and, therefore, an Euler circuit could exist.

21 E

For an Euler circuit to exist, the degree of every vertex in the graph must be even. Only option **E** satisfies this condition.

22 E

For an Euler trail to exist, all vertices except two must have an even degree. Vertices A, B, C and F have odd degree. The only option that connects two of these vertices is A to C, which will leave vertices B and F as vertices of odd degree.

23 D

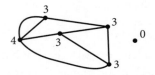

The sum of the degrees is

$4 + 3 + 3 + 3 + 3 + 0 = 16$.

24 C

There are two connections between P to Q (without going through another town). Similarly for P to R and R to Q.

R is the only vertex that has a connection to itself without going through another town.

25 B

Substituting $v = 23$ and $e = 36$ into Euler's formula gives:

$v + f = e + 2$
$23 + f = 36 + 2$
$f = 15$

26 C

Starting at the edge with least weight (2) then next edge (3) from there. Continue choosing edge with least weight until all vertices are included.

minimum spanning tree
$= 2 + 3 + 5 + 4 + 5 + 8 + 8$
$= 35$

9780170465335

27 D

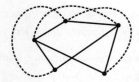

A planar graph has edges that do not cross. It is not possible to 'uncross' the edges of graph **D**.

28 E

The matrix is a 4 × 4 symmetrical matrix, indicating that there are 4 vertices and the graph is undirected. There is one loop on B, two edges connecting C to A and one connecting C to B. This is a total of 4 edges. There is a row and column of zeros in the D position of the matrix, indicating that D is isolated. Drawing a graph will indicate that the sum of the degrees of the vertices is 8.

29 C

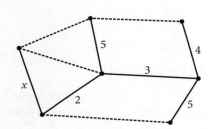

Minimum spanning tree has weight:

$$2 + 3 + 4 + 5 + 5 + x = 24$$
$$19 + x = 24$$
$$x = 5$$

30 B

Shortest length is 22.

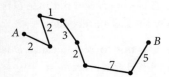

31 C

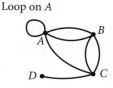

Loop on A

The graph is undirected, so the matrix will be symmetrical about the leading diagonal (which eliminates option **E**). There is one loop on vertex A, which will be represented by a '1' in the A position on the leading diagonal of the matrix and zeros in the other positions (which eliminates options **B** and **D**). There is one edge connecting A to C (which eliminates option **A**).

32 B

An isolated vertex is represented by a row or column of zeros on an adjacency matrix, so this rules out option **A**. In option **C**, vertex D is only connected to itself, so this graph is not connected. In option **D**, vertex A is connected to vertex C, and vertex B to D, but these pairs are not connected. A similar situation exists in option **E**.

33 D

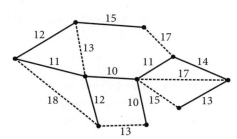

The minimum spanning tree is shown. The length of this minimum spanning tree is 108.

34 B

'The path could have included vertex Q more than once' is the only true statement. Each of the other statements can be demonstrated to be false.

35 E

A strategy to answer this question is to construct a simple graph that satisfies the given conditions and to use this graph to test each option. The graph shown could be used:

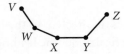

This graph has 5 vertices. Three of even degree (X, Y, W) and two of odd degree (V, Z). From this graph, it can be seen that

- adding an edge between V and Z gives a graph in which all vertices are even and of equal degrees so both options **A** and **B** are possible

- adding an edge between W and Y gives a graph with one vertex of even degree and four of odd degree so option **C** is possible

- adding an edge between V and X gives a graph with three vertices of even degree and two of odd degree so option **D** is possible

- there is no way that a single edge can be added to the graph so that it has four vertices of even degree and one of odd degree so option **E** is correct.

36 C

All the vertices except T and S are of even degree, so an Euler trail (travelling once over each edge) exists, starting at T and finishing at S, that is the sum of all the edges (1340 m). To finish at T, the guard will have to travel ST again, which is an additional 200 m. The total is 1540 m.

37 E

There are only edges coming out of vertex G, so G cannot be reached from any of the other vertices.

38 D

Chantelle plays three sports, whereas Harry, Phil and Michael play two sports.

39 C

Kip only has one-step dominance over Max indicated by the arrow from Kip to Max.

Max has one-step dominance over Lab and Nim, so Kip has two-step dominance over Lab and Nim only.

40 C

Activities on the critical path have their earliest start time (EST) the same as their latest start time (LST).

41 D

The float time of an activity is the difference between the latest start time (LST) and the earliest start time (EST), which in this case is:

$12 - 7 = 5$ hours.

42 B

The critical path is the longest path from the beginning to the end of a project and gives the minimum time for completion of a project. Activities on the critical path cannot be delayed without delaying the whole project.

43 D

The capacity of the cut is:

$5 + 3 + 4 + 4 + 8 = 24$.

The edge with weight '7' is not included because its direction is against the flow from source to sink.

44 D

Joe has driven David, Ben and Vinh to work, but neither Sue nor Claire.

45 A

Process of elimination:

Option **B** is not correct because D is not shown with B as a precedent.

Option **C** is not correct because D is not shown with B as a precedent.

Option **D** is not correct because E is not shown with D as a precedent.

Option **E** is not correct because B is not shown with A as a precedent.

46 C

Activities *A*, *E* and *F* are immediate predecessors of activity *H*. *D* is a predecessor of *F*. *B* is a predecessor of *E* and *D*.

47 C

Forward and backward scanning

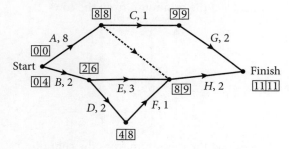

48 A

Activities on the critical path have their earliest start time the same as their latest start time, so *ACG* is the critical path.

49 D

Using the Hungarian algorithm

Step 1

$$\begin{array}{c c c c c} & K & L & M & N \\ W & 3 & 0 & 1 & 3 \\ X & 1 & 0 & 2 & 2 \\ Y & 0 & 2 & 4 & 1 \\ Z & 1 & 0 & 3 & 0 \end{array}$$

Step 2

$$\begin{bmatrix} 3 & 0 & 0 & 3 \\ 1 & 0 & 1 & 2 \\ 0 & 2 & 3 & 1 \\ 1 & 0 & 2 & 0 \end{bmatrix}$$

Step 3

$$\begin{array}{c c c c c} & K & L & M & N \\ W & 3 & 0 & \boxed{0} & 3 \\ X & 1 & \boxed{0} & 1 & 2 \\ Y & \boxed{0} & 2 & 3 & 1 \\ Z & 1 & 0 & 2 & \boxed{0} \end{array}$$

Mei does *W*
Lexie does *X*
Kate does *Y*
Nasim does *Z*.

Minimum time = 4 + 3 + 5 + 2
= 14 minutes

50 A

Removing *Q* to *U* does not isolate a vertex. All vertices still have a path in and out as indicated by the arrows.

51 B

There are seven '1's in the matrix, representing seven edges on the network diagram. Graph *B* is the only one with seven edges and correct directed edges from each vertex as given in the adjacency matrix.

52 C

If *B* is the adjacency matrix, then B^2 will give the number of two-step connections between the vertices.

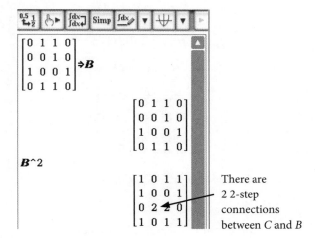

There are 2 2-step connections between *C* and *B*

Or, by observation of the graph, *CAB* and *CDB* are possible.

53 B

A is the only person who can do Task 3. The other allocations are as follow:

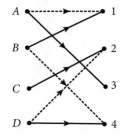

54 D

Apply the Hungarian algorithm:
Subtract the minimum from the rows.

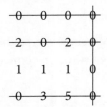

Subtract the minimum from the columns.

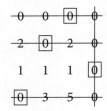

Four lines, so assign the tasks.

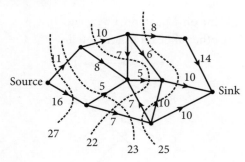

Claire is assigned Task 4.

55 B

maximum flow = minimum cut

Two of the '7's on the minimum cut are against the flow and, are therefore, not counted.

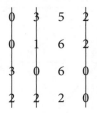

56 B

There is only one loop, on D, which is indicated by a '1' on the leading diagonal of the matrix; this eliminates options **C** and **D**.

From A, there is an edge going to B, C and D, so the first row is: 0 1 1 1.

From B, there is an edge going to C only, so the second row is: 0 0 1 0.

From C, there is an edge going to B and D, so the third row is: 0 1 0 1.

From D, there is an edge going to B and D, so the fourth row is: 0 1 0 1.

57 B

Apply the Hungarian algorithm:
Subtract the minimum from the rows.

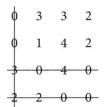

Subtract the minimum from the columns.

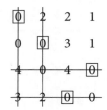

Three lines, so subtract the minimum uncovered, 1, from the uncovered elements and add this number to the elements where two lines intersect.

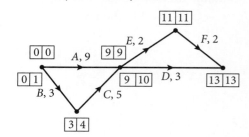

Four lines so assign the tasks.

58 E

The critical path is the longest path through the network. Increasing the time of an activity on a critical path will increase the completion time if there are no other changes.

59 D

Earliest start time and latest start times for the activities are given in the rectangles on the diagram below. The latest start time for activity D is 10 days after the start.

60 C

There are two paths (*AFH* and *BCFH*) of length 18 hours (the critical paths) and two paths (*AEG* and *BCEG*) of length 17 hours. Reducing activities *A* and *B* by 1 hour each reduces the lengths of these two paths to 17 and 16 hours respectively. Reducing either *F* or *H* by 1 hour will then further reduce the critical paths, *AFH* and *BCFH*, to 16 hours. Thus, a minimum of these activities must be reduced by 1 hour each to reduce the project completion time to 16 hours. It is not sufficient to just reduce the length of paths *AFH* and *BCFH* to 16 hours, for example, by reducing the durations of *F* and *H* by 1 hour each because it would still take 17 hours to complete the activities on the paths *AEG* and *BCEG*.

61 D

Earliest start time and latest start times for the activities are given in the rectangles on the diagram below. The float time of an activity is the difference between the latest start time and the earliest start time of that activity. For activity *E*, this is $11 - 7 = 4$ days.

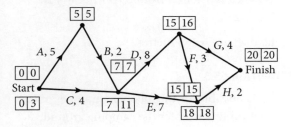

62 A

One of the '6's on the minimum cut is against the flow and, therefore, is not counted.

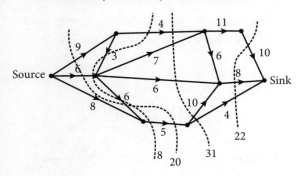

63 D

The critical path for a project is the longest path from start to finish. Using forward and backwards scanning, the critical path is the path where $\boxed{EST \mid LST}$ is the same.

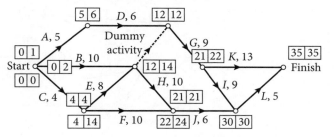

Extended-answer solutions

64 Brendan's cuts

Question 64.1

a 3 and 2

b maximum flow = 5

c

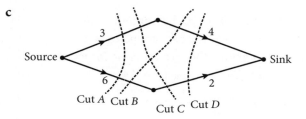

d The capacity is the sum of the flows that are from source to sink across the cut.

Cut *A* has capacity 9.
Cut *B* has capacity 10.
Cut *C* has capacity 5.
Cut *D* has capacity 6.

e Cut *C*, with the maximum flow of 5, is consistent with the prediction in part **b**.

Question 64.2

a Minimum cut gives:

maximum flow = 2 + 4 + 3 + 5 = 14

b Cut 1 does not entirely stop the flow from source to sink.

c Capacity of cut 2 = 3 + 1 + 2 + 6 + 3 + 5 = 20

d The first 2 is not counted because the flow is from right to left.

9780170465335

e   Minimum cut drawn as shown:

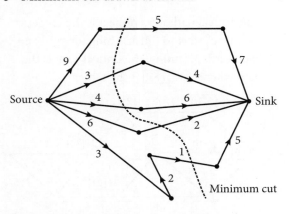

Maximum flow = 1 + 2 + 4 + 3 + 5 = 15

Question 64.3

a   Capacity of Cut Q is:

$x + 2 + 4 + 3 + y = 15$, i.e., $x + y = 6$

All combinations of x and y are:

0 and 6, 1 and 5, 2 and 4, 3 and 3, 4 and 2, 5 and 1, 6 and 0.

b   The capacity of the minimum cut has previously been found to be 15. Therefore, the greatest value for y will give Cut P a capacity of 14:

capacity of Cut P = 1 + 2 + 4 + 3 + y = 14

Solving for y gives: $y = 4$

65 Camping

Question 65.1

a   Yes, each vertex is of even degree, so if you travel along one edge into the vertex, then there will be another edge to travel out on.

b   This is called an Euler circuit.

Question 65.2

a   Start at 'Office', then *AEIJHGFBCD*, then 'Office'. There are many possibilities; each of the 10 letters should be mentioned once only.

b   Hamiltonian cycle

Question 65.3

a   Below is one possibility. A spanning tree is a connected graph with no circuits.

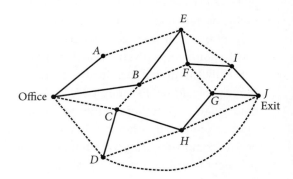

b

Question 65.4

The shortest distance from the Office to J is: Office *BFIJ*.

total distance = 185 metres

Question 65.5

a   It is possible for the following number of cars per minute to enter the camping ground:

$5 + 7 + 5 + 10 = 27$

This is the sum of the capacities of the edges leaving the Office.

b   Maximum flow through the camping ground is the capacity of the minimum cut, which is 24.

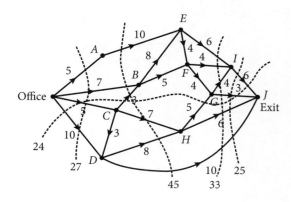

Question 65.6

Only Greg wants to collect rubbish, so he is allocated this task. Jess, Nancy and Bill are then allocated the only tasks available to them. Alphonse and Maria have chosen the same tasks, so there are two possible allocations for them: Alphonse to clean showers and Maria to clean campsites, or vice versa.

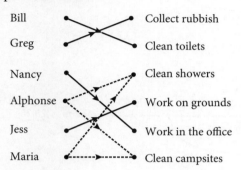

66 Crashing by design

Question 66.1

a

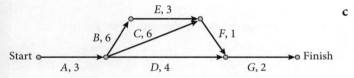

Activity C with Activity A as predecessor, and Activity E with Activity B as predecessor.

Activity F with activities C and E as predecessors, and Activity G with activities D and F as predecessors.

All arrows are included and times labelled.

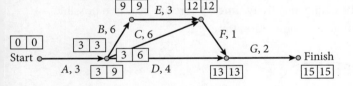

b **i** $A \to B \to E \to F \to G$

ii These are the activities that cannot be delayed if the renovations to the salon are to be completed in minimum time.

c Earliest start time for Activity F is 12 weeks. Latest start time for Activity C is 6 weeks.

d 15 weeks

e As Activity C is a non-critical activity, reducing only its time will not alter the minimum time for completion.

Question 66.2

a Even if all activities that may be reduced are reduced as much as possible, the minimum completion time cannot be less than 11 weeks (4 weeks less than the original).

b Cost of reducing by 3 weeks is $9000.

This is achieved by reducing Activity E by 1 week and Activity B by 2 weeks

i.e. $1 \times 1000 + 2 \times 4000$.

The activities to be reduced to save 4 weeks are E, B and C.

Activity B may be reduced by a further week, to 3 weeks, but this also means Activity C must be reduced by 1 week, otherwise Activity C would take one more week than the combined totals of activities B and E and the minimum time would be 12 weeks along a new critical path of:

$A \to C \to F \to G$

Cost of reducing by 4 weeks = $16 000

c

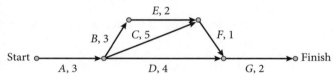

The network as before, but with new correct times for activities B, C and E.

d There are now two critical paths, with both having a minimum time of 11 weeks:

$A \to B \to E \to F \to G$

and:

$A \to C \to F \to G$

67 Landscape

Question 67.1

a

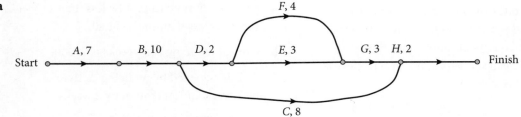

b

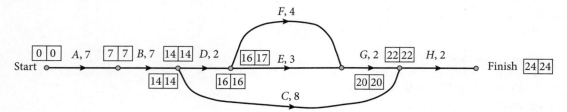

The maximum reduction in time will be 4 days, from 28 days to 24 days. The total cost of this reduction will be $1500.

c The critical path includes only the activities that have the same values for their earliest start time and latest start time. So the critical path for this project is *ABDFGH*.

d The minimum completion time is 28 days.

e The float time is the difference between the earliest start time and the latest start time. For Activity *E*, this is:

$20 - 19 = 1$ day

Question 67.2

a To reduce the overall completion time, tasks on the critical path need to be reduced. Only activities *B* and *G* (of those possible to reduce) are on the critical path. Reducing Activity *B* will not affect any of the other activities, so Activity *B* can be reduced by 3 days. Reducing Activity *G* by 1 day will also not affect any of the other activities, so Activity *G* can also be reduced by 1 day.

total cost = $3 \times \$400 + 1 \times \$300 = \$1500$

Below is the forward and backward scan for the changed activity times. There are now two possible critical paths. Reducing Activity *E* will not change the completion time.

b If only a maximum of $1000 is to be spent, then this can be done by either reducing Activity *B* by 2 days or activities *B* and *G* by 1 day each. Reducing Activity *B* by 2 days costs $2 \times \$400 = \$800$ and reducing Activity *B* by 1 day and Activity *G* by 1 day costs $\$400 + \$300 = \$700$.

Both options reduce the overall completion time by 2 days and the cheaper option should be taken. The minimum finishing time is 26 days.

Question 67.3

a We are looking for the minimum spanning tree.

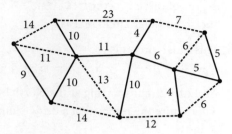

b Adding the weightings on the edges of the minimum length spanning tree gives 74 metres.

68 Neighbours

Question 68.1

a

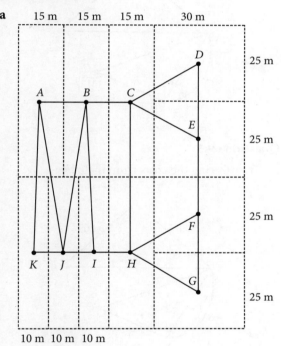

b

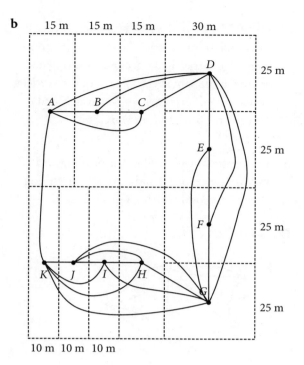

Question 68.2

A minimum spanning tree:

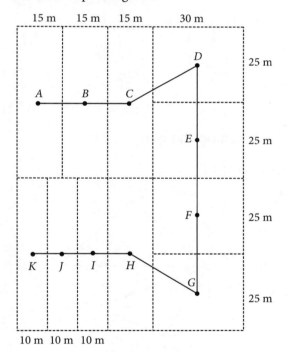

Question 68.3

a To complete an Euler circuit (which means each edge is travelled once only, starting and ending at the same vertex) each of the vertices needs to be of even degree. Vertices N and L have degree 3, so an Euler circuit is not possible.

b The postman will need to travel along NM and ML twice, or NI and IL twice. These edges connect the odd vertices at N and L.

Question 68.4

a

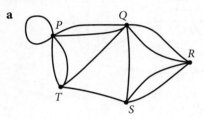

b

$$
\begin{array}{c}
 \\
\begin{array}{c} P \\ Q \\ R \\ S \\ T \end{array}
\end{array}
\begin{array}{ccccc}
P & Q & R & S & T
\end{array}
\left[
\begin{array}{ccccc}
1 & 2 & 0 & 0 & 2 \\
2 & 0 & 2 & 1 & 1 \\
0 & 2 & 0 & 2 & 0 \\
0 & 1 & 2 & 0 & 1 \\
2 & 1 & 0 & 1 & 0
\end{array}
\right]
$$

69 Bridging the gap

Question 69.1

a

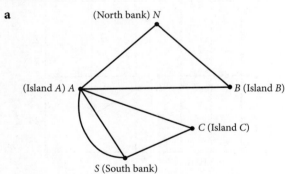

b The degree of A is 5 because there are five edges attached to A.

c This graph does not contain an Euler circuit (which is a circuit travelling each of the edges once, starting and finishing at the same vertex). An Euler circuit only exists when the degree of each of the vertices is even. This graph has odd vertices at A and S.

d i An Euler circuit starting at the North bank will require an additional bridge joining A and S. This will make the degree of these two vertices even and so an Euler circuit exists.

ii This Euler circuit is
$N \to B \to A \to C \to S \to A \to S \to A \to N$.

e i An Euler circuit starting at the North bank will require the removal of a bridge joining A and S.

ii This Euler circuit is
$N \to B \to A \to C \to S \to A \to N$.

Question 69.2

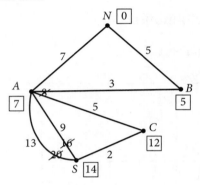

Question 69.3

a Transferring the costs from the table:

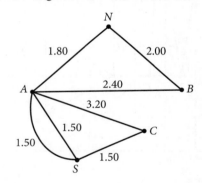

b

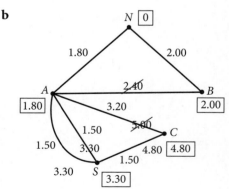

c The minimum cost will be the path
$N \rightarrow B \rightarrow A \rightarrow S \rightarrow C \rightarrow S$ with a cost of

$2.00 + 2.40 + 1.50 + 1.50 + 1.50 = \$8.90$.

70 La Principessa

Question 70.1

a

Task	Predecessors
A	–
B	–
C	A
D	B
E	C, D

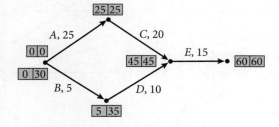

b Tasks *A*, *C*, and *E* are on the critical path.

c 60 minutes

d Tasks *B* and *D*, because they are not on the critical path.

e Task *B* has a latest starting time (LST) of 30 minutes. Task *D* has an earliest starting time (EST) of 5 minutes. Task *D* has an LST of 35 minutes.

f The float time is the difference between LST and EST, so Task *B* and Task *D* both have a float time of 30 minutes.

Either task could be delayed for 30 minutes without affecting the minimum time for project completion.

Question 70.2

Stephen to Supplier 3, Peter to Supplier 2, Michelle to Supplier 1, Monica to Supplier 4.

The steps for the working of the Hungarian algorithm are given below.

Step one

Subtract the minimum entry in each row from each element in the row.

	Supplier 1	Supplier 2	Supplier 3	Supplier 4
Stephen	5	0	11	3
Peter	5	0	13	3
Michelle	0	2	18	0
Monica	3	4	9	0

Step two

Repeat for any column without a '0' entry.

	Supplier 1	Supplier 2	Supplier 3	Supplier 4
Stephen	5	0	2	3
Peter	5	0	4	3
Michelle	0	2	9	0
Monica	3	4	0	0

Step three

Minimum number of lines to cover the zero elements is three (through column 2, and rows 3 and 4). Since this is less than the number of rows, proceed to Step 4.

Step four

The lowest uncovered element is 2. Add this to the elements covered by 2 lines and subtract it from each uncovered element.

	Supplier 1	Supplier 2	Supplier 3	Supplier 4
Stephen	3	0	0	1
Peter	3	0	2	1
Michelle	0	0	9	0
Monica	4	6	0	0

Step five

Cover the zeros with the minimum number of lines. Four lines are now needed so the tasks can be allocated. Allocate the tasks. Look for the rows and/or columns that have only one zero.

	Supplier 1	Supplier 2	Supplier 3	Supplier 4
Stephen	-------3-------	-------0-------	-------[0]-------	-------1-------
Peter	-------3-------	-------[0]-------	-------2-------	-------1-------
Michelle	-------[0]-------	-------4-------	-------9-------	-------0-------
Monica	-------3-------	-------6-------	-------0-------	-------[0]-------

Step six

The minimum number of lines needed to cover the zeros is now four, so a bipartite graph can be drawn with the edges chosen from the zero elements.

From this graph, Peter must go to Supplier 2. This leaves Stephen with Supplier 3 as his only option. This, in turn, leaves Monica with Supplier 4 as her only option and Michelle with Supplier 1 as her only option.

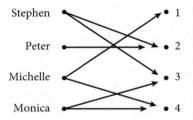

71 Security

Question 71.1

a *ABCDBFEDFAE* or *AEFBCDBAFDE* (others possible)

b We are being asked for an Euler circuit (which means travel along each edge once only) and this is only possible if the degree of each of the vertices is even. Vertices *A* and *E* are of odd degree, so an Euler circuit is not possible. (An Euler trail is possible if it starts and ends at vertices of odd degree.)

c The path from *A* to *E* (or *E* to *A*) must be travelled twice (in opposite directions) to start and end at vertex *A*.

d The minimum distance starting and ending at vertex *A* will be the sum of the edges plus the distance *AE*.

$AE = 80$ m

total distance $= 6 \times 50 + 2 \times 60 + 2 \times 80 + 80$
$\qquad\qquad\quad = 660$ metres

Question 71.2

a *PQRSTUVWP* or *PVUTSRQWP* (others are possible, in which nine vertices are listed with only *P* mentioned twice [at the start and the end]).

b The route *PVUTSRQWP* has length 609 metres.

Question 71.3

a Use the Hungarian algorithm:

$$\begin{bmatrix} 8 & 12 & 18 & 13 \\ 7 & 11 & 16 & 12 \\ 12 & 11 & 17 & 15 \\ 8 & 13 & 15 & 17 \end{bmatrix}$$

Subtract the minimum from each of the rows.

Only two lines are needed to cover the zeros.

$$\begin{bmatrix} 0 & 4 & 10 & 5 \\ 0 & 4 & 9 & 5 \\ 1 & 0 & 6 & 4 \\ 0 & 5 & 7 & 9 \end{bmatrix}$$

Subtract the minimum from each of the columns.

$$\begin{bmatrix} 0 & 4 & 4 & 1 \\ 0 & 4 & 3 & 1 \\ 1 & 0 & 0 & 0 \\ 0 & 5 & 1 & 5 \end{bmatrix}$$

Only two lines are needed to cover the zeros.

Subtract the minimum uncovered element (1) from each of the uncovered elements and add it to the elements that are covered by two lines.

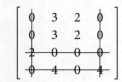

Four lines are needed to cover the zeros.

The guards can be allocated to the locations:

$$
\begin{array}{c}
 & A & B & C & D \\
\text{Theo} \\
\text{Reg} \\
\text{Tanya} \\
\text{Jack}
\end{array}
\begin{bmatrix}
0 & 3 & 2 & 0 \\
0 & 3 & 2 & 0 \\
2 & \boxed{0} & 0 & 0 \\
0 & 4 & \boxed{0} & 4
\end{bmatrix}
$$

Tanya can go to location B and Jack can go to location C. There are now two remaining possible allocations of guards (Theo and Reg).

Theo to location A and Reg to location D or Theo to location D and Reg to location A.

b The minimum total distance travelled by the guards is:

$11 + 15 + 8 + 12 = 46\,\text{km}$
(or $11 + 15 + 13 + 7 = 46\,\text{km}$).

72 VCAA 2010 Exam 2

Question 72.1

a No communication is permitted between the two members.

b $f = 1, g = 0$

Question 72.2

a 11 minutes

b **i** Hamiltonian path

ii

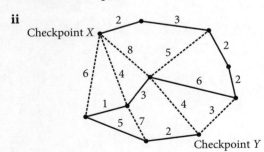

Question 72.3

a D

b **i** B, D, F

   ii $32\,\text{km}$

c B and D

Question 72.4

a 2   **b** 9 minutes

c A, C **d** 4 minutes

e 16 minutes   **f** $ABDH$

73 VCAA 2011 Exam 2

Question 73.1

a $200\,\text{km}$

b $FDC, FEDC, FEBC, FEABC, FDEBC, FDEABC$
$\Rightarrow 6$

c Bredon

d $240\,\text{km}$

Question 73.2

a

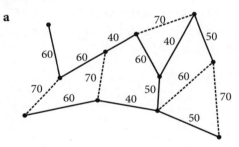

b $510\,\text{m}$

Question 73.3

a 2 hours

b 3 hours

c F, H (both have float time of 3 hours)

d 13 hours

e 14 hours

Question 73.4

a Stormwater from Source 2 cannot reach Outlet 1.

b Outlet 1 = 700 kilolitres per minute

   Outlet 2 = 700 kilolitres per minute

c 300 kilolitres per minute

SOLUTIONS – CHAPTER 4

74 VCAA 2012 Exam 2

Question 74.1

a   **i** 160 m

    ii 2

    iii 1250 m

-b **i**

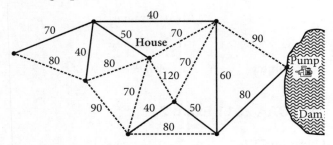

    ii minimum spanning tree

Question 74.2

a 12 days

b Activity *F* has only activity *B* as a predecessor, while activities *G* and *H* have both *B* and *C* as predecessors. As there cannot be two activities called *B*, a dummy activity (with zero time) is drawn as a form of extension of *B* to the start of *G* and *H* to indicate that *B* is a predecessor for these two activities as well.

c 15 days

d *ABHILM* (28 days)

e 25 days

Question 74.3

a

Task	Worker			
	Julia	**Ken**	**Lana**	**Max**
W	5	0	1	4
X	10	5	0	17
Y	9	6	0	7
Z	12	0	0	9

b A minimum of 4 lines needed (as 4 tasks to be allocated).

c

Task	Worker			
	Julia	Ken	Lana	Max
W	0	0	4	0
X	2	2	0	10
Y	1	3	0	0
Z	7	0	3	5

d

Worker	Task
Julia	*W*
Ken	*Z*
Lana	*X*
Max	*Y*

75 VCAA 2014 Exam 2

Question 75.1

a 2

b miniature trains

Question 75.2

a

Task	Andrew	Brianna	Charlie	Devi
Publicity	3	2	0	0
Finances	0	1	2	2
Equipment	0	4	3	2
Catering	1	2	3	0

b Minimum number of lines to cover zeros is 3. This is less than the number of rows.

Task	Andrew	Brianna	Charlie	Devi
Publicity	4	2	0	1
Finances	0	0	1	2
Equipment	0	3	2	2
Catering	1	1	2	0

c **i** equipment

ii 36 hours

Question 75.3

a **i** Bower, Eden

ii 910 km

b 270 km; Bower – Clement – Derrin – Eden

c between Bower and Derrin

Question 75.4

a 7 hours   **b** 18 hours   **c** 2 hours

d 4 hours   **e** $270

Formula sheet

General Mathematics exams 1 & 2

© Victorian Curriculum and Assessment Authority, 2023

Data analysis

standardised score $\qquad z = \dfrac{x - \bar{x}}{s_x}$

lower and upper fence in a boxplot $\qquad$ lower $\quad$ Q1 $- 1.5 \times$ IQR $\qquad$ upper $\quad$ Q3 $+ 1.5 \times$ IQR

least squares line of best fit $\qquad y = a + bx, \qquad$ where $b = r\dfrac{s_y}{s_x} \qquad$ and $\qquad a = \bar{y} - b\bar{x}$

residual value $\qquad$ residual value = actual value − predicted value

seasonal index $\qquad$ seasonal index $= \dfrac{\text{actual figure}}{\text{deseasonalised figure}}$

Recursion and financial modelling

first-order linear recurrence relation $\qquad u_0 = a, \qquad u_{n+1} = Ru_n + d$

effective rate of interest for a compound
interest loan or investment $\qquad r_{effective} = \left[\left(1 + \dfrac{r}{100n}\right)^n - 1\right] \times 100\%$

Matrices

determinant of a 2 × 2 matrix $\qquad A = \begin{bmatrix} a & b \\ c & d \end{bmatrix}, \qquad \det A = \begin{vmatrix} a & b \\ c & d \end{vmatrix} = ad - bc$

inverse of a 2 × 2 matrix $\qquad A^{-1} = \dfrac{1}{\det A}\begin{bmatrix} d & -b \\ -c & a \end{bmatrix}, \qquad$ where $\qquad \det A \neq 0$

recurrence relation $\qquad S_0 =$ initial state, $\qquad S_{n+1} = T S_n + B$

Leslie matrix recurrence relation $\qquad S_0 =$ initial state, $\qquad S_{n+1} = L S_n$

Networks and decision mathematics

Euler's formula $\qquad v + f = e + 2$